The Misunderstood Universe:

Correcting and explaining the cosmic
mistakes, mysteries, and beliefs,
including Dark Matter, Dark Energy,
pioneer anomaly, age of the universe,
Hubble red shift due to gravity,
quasar velocity, tired light,
the expanding universe, inflation,
the cosmic microwave background,
black holes, and the Big Bang

The Misunderstood Universe:

Correcting and explaining the cosmic
mistakes, mysteries, and beliefs,
including Dark Matter, Dark Energy,
pioneer anomaly, age of the universe,
Hubble red shift due to gravity,
quasar velocity, tired light,
the expanding universe, inflation,
the cosmic microwave background,
black holes, and the Big Bang

By Sol Aisenberg, PhD

iUniverse, Inc.
New York   Bloomington

The Misunderstood Universe:

*Correcting and explaining the cosmic mistakes, mysteries, and beliefs,
including Dark Matter, Dark Energy, pioneer anomaly, age of the universe,
Hubble red shift due to gravity, quasar velocity, tired light, the expanding universe,
inflation, the cosmic microwave background, black holes, and the Big Bang*

iUniverse books may be ordered through booksellers or by contacting:

*iUniverse
1663 Liberty Drive
Bloomington, IN 47403
www.iuniverse.com
1-800-Authors (1-800-288-4677)*

*Because of the dynamic nature of the Internet, any Web addresses or links contained in this
book may have changed since publication and may no longer be valid. The views expressed
in this work are solely those of the author and do not necessarily reflect the views of the
publisher, and the publisher hereby disclaims any responsibility for them.*

*ISBN: 978-1-4401-9179-4 (sc)
ISBN: 978-1-4401-9181-7 (hc)
ISBN: 978-1-4401-9180-0 (ebook)*

Printed in the United States of America

Library of Congress Control Number: 2009911967

iUniverse rev. date: 12/30/09

Acknowledgements

This material is dedicated to my lovely, smart, and patient wife, Ruth who watched me work and accepted my obsession.

Other incentives included my children, Anne, Mark, and Amy, as well as my grandchildren, Lily and Serena.

The work of past and current generations of scientists, engineers and others interested in the universe is gratefully acknowledged. They provided important observations and theories to study, evaluate, and extend.

Preface

This information and analysis about the misunderstandings and mistakes in the standard model of the universe is one of my contributions to the current and future scientific community.

I expect that many of my contributions will be rejected by some of the experts in the field of cosmology, and this is understood because it may question the prior work of experts. However, I have read that progress in science progresses "funeral by funeral" and then the next generations take over without ownership or commitment to some prior ideas.

Starting in early 1900, with the increased availability of observations of the remote stars and galaxies at cosmic distances, a number of mysteries appeared in the supposed understandings of the observations. They apparently continue to the present day

In various portions of this document, I will discuss the different aspects of the Misunderstood Universe, as I see it, along with my suggested corrected views of the mysteries in the universe.

As a trained scientist and physicist experienced in many fields, starting in about 1998, I learned of the major mysteries, such as Dark Matter and Dark Energy, in the standard model of the universe. There was curiosity and a need to understand the mysteries. While working on some of my patent applications and inventions, and other scientific problems, along with my consulting company, I obsessively and part time also considered solutions of the mysteries and studied the supposed support for these mysteries.

The result is that I determined that the problem was due to an incomplete understanding, by others, of the key role of gravity in the model of the universe.

The published and careful observations by Vera Rubin of the flat rotation velocity curves in spiral galaxies led me to question the assumption that Newton's law of gravity is a universal law. The genius Newton was only able to base his theory on available observations in our solar system and not at cosmic distances where information was only available later.

I was able to arrive at a beautifully simple extension of Newton's gravitational constant that removes the need for dark matter, and that unexpectedly also explains the role of gravity in the redshifts and removed the need for dark energy, It also explained many of the other conflicts and speculations in the standard model of the universe. These will be described in more detail in subsequent chapters.

My theory of additional gravity (TAG) to be presented in later chapters is an addition to the Newton and Einstein's theories of gravity, and must not be considered as a removal or replacement. It appears to fit the observations and to remove the need for the long-term unsuccessful search for massive amounts of missing matter or Dark Matter

The early gravity related contributions of Albert Einstein related to the effect of gravity on photons (his General Relativity) and the contribution of Fritz Zwicky related to the effect of gravitational drag on on photons and the red shift (tired light) were important in correcting the standard model of the universe, and in understanding the vital role of gravity in the universe.

In spite of the frequent statements that gravity is the weakest force, compared to electrical and magnetic forces, and nuclear forces it should be noted that only gravitational forces are sufficient to collect large amounts of interstellar gas and to provide the gravitational potential energy needed for the ignition of stars during gravitational collapse of collections of gas, thus providing the density and temperature needed to start nuclear processes in stars, and the subsequent release of higher atomic weight elements.

Gravity plays an important role in influencing the red shift in the photons conveying information about the cosmos to observers.

The recent and continuing referrals to Dark Matter and Dark Energy and other mysteries in the scientific literature, magazines, and in television programs convinced me that society needs to have a true understanding of the universe, without the fascinating and misleading mysteries.

The very recent introduction of the mystery of Dark Energy, in 1998, was based upon published observations of Type Ia supernovae and suggested that the supposed expansion of the universe was accelerating. This led to the reexamination of the observational work of Hubble related to the red shift. However, the interpretation of the red shift observations and the assignment of them to the "apparent Doppler effect" have not been proven. The result is the unproven belief that the red shift showed the receding velocity of the observed supernovae and apparently indicated that the universe is expanding.

Careful examination of the implications of the linear dependence of the red shift shows a logical error in the belief. There is a limitation in the extrapolation to very large distances. The red shift corresponds to the loss of photon energy with travel, and if one considers that the effect of far travel distances it would eventually require the impossibility of the need for the photon energy to become zero and eventually negative.

I have presented various comments about my evolving understanding of the misunderstood universe on the Internet in the past ten years, and if you are interested, you can probably scc them by a Google search on ["Sol Aisenberg" gravity], and omit the brackets.

If my analysis and suggested corrections are accepted by this present generation, it will help redirect the work of many excellent scientists, physicists, and engineers, and help produce an improved understanding of the universe.

Some Comments

Investment of time and money in science and technology is importation for the growth and survival of society, in the past and in the future. Without mankind's great curiosity, we would be living under primitive conditions. Investment in science has brought great returns - much greater than for many other investments. Unfortunately the recent reduced investment in science and research has not been sufficient. In one sense we are eating our seed corn and our future dependence on the results of diminished research and science will be disappointing and dangerous if not promptly corrected.

Some areas of research and curiosity may not be immediately productive, but the study of the universe will help improve our experimental work in many other areas, and will train new scientists who may use their training in areas with more immediate benefits.

We should acknowledge that what we learn about the universe probably would not have any practical use in the immediate future, but we should insure that our observations should not be devalued by the mistaken theories that result in mysteries such as Dark Matter and Dark Energy, and unsupported belief in the Big Bang. Learning how to solve problems correctly and with an open mind is an important tool for future scientists.

Unless we or other species learn how to travel faster than light we probably will never have access to other regions in the far universe.

Please excuse my apparent lecturing, but I feel strongly that the message about support of research is important.

With respect to the understanding of the universe, there will remain other tasks for those interested in this field, including refinement of my contributions.

In any event, the information provide here can be of help to future generations.

In our opinion and probably that of others, the valid observations must take priority over theory when the theory is not in agreement with observations.

Speculation can be the beginning of understanding but should not exist in valid solutions or explanations.

An indicator of speculations can include any of the following words, for example: apparently, suggests, believe, suppose, maybe, perhaps, if, maybe, - or similar phrases.

My personal belief is that science is based upon observations. Whenever the theory does not agree with observations (validated by repetition, and preferably by others) then the observations should take priority. The speculation or theory in order to be considered must be modified or extended so as to agree with the observations so that it can be considered valid. Theories however, have value because they can guide the attempted validation by new observations.

An acceptable theory for consideration is one that can agree with past observations, and even more important can suggest and agree with future observations.

My observation of the conflicts in the standard model of the universe have convinced me that the current work on the understanding of the universe is on the wrong track, and must be corrected.

The funded work of some of the excellent scientists, physicists, and engineers in this field should be redirected to more practical tasks in solving problems, such as but not limited to providing (a) new massive and clean energy sources, (b) solving impending shortage of clean water, (c) removing the limitation of food supplies, and (d) improving the health of people worldwide.

Humanity will need vast amounts of energy to support the existence of (perhaps) reduced populations during future periods of centuries of ice ages and increased temperatures.

Sol Aisenberg

Contents

Introduction

This analysis of the standard model of the universe will analyze, identify, warn, and explain the various reasons and ways that the standard model of the universe is seriously misunderstood and misleading.

Much of the information provided in the various chapters will repeat material in other chapters because this is a collection of essays, chapters that were prepared over periods of time, and were expanded using new insights and used different ways of explaining my analysis.

It is probably useful in repeating the important material in alternate ways to insure that the reader can absorb the material and new concepts.

Also, it may not be necessary to read and study all the chapters, but the reader can focus on the chapters of particular concern before perhaps studying all the material.

The information can be read and understood by those without knowledge of mathematics or equations. Only a few simple equations were included for the benefit of scientists to help explain and support my arguments.

Some of the chapters are smaller than others because it focuses on one idea, and because it provides a change of pace. Additional details of the idea can be found in other chapters.

As a result of my contribution, if and when my analysis is accepted, much of the past theoretical work of many experts, scientists, and physicists may be shown to need corrections and be of less value. However, the equipment built and operated by excellent engineers and scientists to add to the collection of observations will remain of value, and can help the future theoretical work to repair the standard model of the universe and to lead to a true understanding the universe.

As a scientist I value and use theory to guide my experimental work, but when there is a conflict between valid observations and theory, the observations must take priority over theory, and the theory must be reexamined and corrected.

When a theory and model contains singularities, and speculations, and mysteries and massive fudge factors such as Dark Matter and Dark Energy, and is not supported by observations, it is evidence to me that the theory and model are just unsupported speculations and must be abandoned or at least refined and corrected.

In the last 100 years, starting with the increased information about other galaxies outside our own galaxy, the Milky Way, our observations and supposed knowledge of the universe expanded, but introduced many mysteries such as Dark Matter and Dark Energy.

The problems and misunderstandings began with the important work of Hubble in the 1900s that identified the presence of remote galaxies and measured their cosmic distances associated red shifts. These observed red shifts together with the distances of the galaxies as determined by various observations (including parallaxes), showed an apparent linear relationship between the distances and the red shifts.

When Hubble, ascribed the red shifts to "… an apparent Doppler effect …" this started a belief that the galaxies were receding, which resulted in a belief that the universe was expanding. However, it should be realized and admitted that there are not any direct measurements or observations of the receding velocity to support this belief, and speculation.

The incorrect belief in the receding velocity of galaxies is responsible for many of the mistaken theories of the universe, including the wrong age of the universe and the belief in the big bang, among others.

Details of the various mistakes and mysteries, and the correct views are discussed in more detail in subsequent chapters.

Actually all of the various aspects of the universe are based upon gravity, which itself is the ultimate mystery. Gravity can be seen only through its effect on observable entities.

In order to prevent my contribution to be accused as just speculation, I will base it only on observations of others, and will try to point out where the existing model is itself just based upon speculations.

As an example, there is speculation that Newton's law of gravity is a universal law, although Newton did not know much about galaxies outside our local area.

Another example is the belief and very harmful speculation that the red shift measured by Hubble can be used to demonstrate the receding galaxies.

I have been a member or the American Physical Society since undergraduate days, and have been a reviewer of many articles for them. When I twice recently tried to submit articles clarifying various aspects of the mistaken beliefs of the standard model of the universe, they were rejected (apparently without going to reviewers and without reviewer comments) and were termed just speculation by the editor. Some parts of my information were presented and published as abstracts of poster sessions and brief talks for regional meetings of the American Physical Society where reviewers were not involved. If you want to do a Google search on ("sol aisenberg" gravity) you can find my earlier postings although their contents are expanded and included in several of the chapters of this book.

Recognizing, as a problem solver, that it would be a waste of time trying to introduce new ideas this way because of establishment resistance and inertia, I decided to write and publish this book, and share some free copies to reviewers and others who might take a more professional attitude. The first edition of this book, "The Misunderstood Universe" is the result and a second edition may be prepared if it appears suitable. I can now spend more time in my inventing activities, including important energy solutions, medical solutions, security improvements, and other immediate needs of society.

The understanding of the universe was considered and advanced in stages by many people in past centuries. Notable were Aristotle (340 B.C.), Ptolemy, Copernicus (1514), Kepler, Galileo (1609), Newton (1687), Olbers (1823), Maxwell (1865), Einstein (1905), and Hubble (1929). Note that the differences in dates approximately represent successive generations. It is sometimes said that advances progress funeral by funerals (apparently the supporters of the prior objections and beliefs must disappear for the new ideas to flourish).

Some of the chapters contain information that may be of particular interest to readers although the some of the details are redundant but explain in different ways. They include Chapters 1,2, 3, 4, 5, 6, 7, 13, 14, 21, 25, 26, 27, 28, 34, and 35, because the involve the more important aspects of the misunderstood universe.

For additional information, you may contact me through:

Tel: 508/651-0140
solaisenberg@comcast.net
saisenberg@alum.mit.edu

Abstract

The Misunderstood Standard Model Of The Universe
Sol Aisenberg
International Technology Group
solaisenberg@comcast.net   saisenberg@alum.mit.edu

There are two fundamental, serious beliefs, errors, and assumptions in the standard model of the universe, which are responsible for the current misunderstood universe. Theyresult in the apparent need for mysterious massive Dark Matter and Dark Energy. The first assumption is that Newton's law of gravity is a universal law, and that it is also valid at cosmic distances, and not just in our solar system, (it is based upon observations of motion of planets in our solar system). When applied to observations of motion of galaxies at larger distances, Newton's law fails to explain these observations. Under the current model it requires assumptions of massive amounts of invisible missing matter or Dark Matter to supply the necessary gravity. We find that our simple Theory of Additional Gravity (TAG) as an addition to Newton's original law of gravity for large distances will fit the observations at cosmic distances and also apply, without change, in our solar system. Our TAG model adds a linear term to Newton's gravitational constant and this term results directly from an elementary equation for the balance between centrifugal and gravitational forces in spiral galaxies. The equation, $M*G=r*v*v$, requires that $M*G$ is a liner function of distance, r, at distances where the velocity, v, is observed to be constant. This theory is actually different from the interesting MOND theory of M. Milgrom, which is related to acceleration.

The second assumption and error, with much more serious consequences, is the belief that Hubble's law always describes red shifts as a linear function of distance, and that it is due to the Doppler effect. This leads to the belief that the stars and galaxies are receding, and also that the red shift can accurately reveal distances even for very remote galaxies. It apparently showed that because the galaxies are receding, the universe is expanding, and that there was a big bang. Actually, velocities were not directly observed and thus this is not evidence of the expansion. Also the age of the universe deduced from the Hubble constant is wrong. In fact, the Hubble red shift law cannot even be used for distance determination where extrapolation of the linear dependence on distance to very large distances would require an unlikely zero and even negative photon energy. Thus for observations of very remote galaxies, the difference between the star distances determined by observations of star magnitudes, the distances from observed red shifts apparently wrongly showed acceleration of the supposed expansion and the need for vast Dark Energy. Tired light due to energy drain by gravity is explained along with the features of the Cosmic Microwave Background, CMB, and its low temperature.

Chapter 1

Summary Of The Most Important Conclusions

In the following portions of this document I will explain the reasons for the conclusions I reached after about ten years of, part time, obsessive, investigation (as a curious scientist) of the many mysteries in the supposed understanding of the universe. To guide the reader, I will identify the most important conclusions, and then the associated consequences will be discussed later in the document.

Gravity, even if the source or cause is not understood, is important and vital in most events and processes in the universe.

Without gravity, matter in the universe would not collect into masses such as stars, and galaxies. Gravitational forces cause the collapse of material in stars thus igniting the stars through gravitational energy produced by the gravitational collapse. This also causes the start of nuclear processes resulting from fusion associated with high temperature, and high density and gravitational confinement.

The nuclear processes in stars produce star explosions and the release of the higher atomic mass materials that make up part of the universe. Gravity also influences the energy of photons emitted from stars, galaxies, quasars, and also drains energy from photons traveling large distances, resulting in tired light.

There is no need for the invisible Dark Matter proposed to provide the necessary gravitational forces to explain the puzzling observations at cosmic distances and the related Dark Matter. Actually, the observations for spiral galaxies indicate a questionable linear invisible addition to the mass, or else a linear invisible addition to the already invisible gravitational constant. Because gravity itself is invisible, we show that when an invisible linear additional gravity term is added to the invisible Newton's gravity, the result is in agreement with the associated observations, thus eliminating the need for invisible Dark Matter.

Second, there is no need for Dark Energy. The apparent need is because of a belief that the red shifts and photon energy loss described by the Hubble observations for nearby cases show the red shift as a linear function of distance plus the assumption that it can be extended linearly to measure very large distances. However, for very, very large distances, the assumption that the linear form is also valid for extreme distances would eventually require the photon energy to go to zero and then even negative – not very likely.

Third, the red shifts are not mostly due to the Doppler effect, and do not prove receding galaxies, or an expanding universe.

Fourth, the red shifts have large contributions due to gravity draining photon energy while traveling, in addition to any smaller Doppler effects.

Fifth, there is an alternate explanation for the cosmic microwave background, CMB, radiation other than as the remains of the big bang, and also one for the low 2.7 K degree temperature of the radiation.

Sixth, there was no proof of the big bang because there are no valid direct observations or demonstrations of the receding galaxies or of the expanding universe.

Seventh, tired light as an explanation of the Hubble linear dependence of the red shift on distance can be understood and predicted as a result of the process involving the drain of photon energy by gravitational interaction with interstellar gas and dust without collisions - similar to gravitational tidal interaction and transfer of energy from our moon to earth.

There are many more explanations of other and less significant conflicts in the standard model of the universe, and they will be covered in other chapters.

As an example of the wrong reasoning in the understanding of the universe, consider the illogical proof of the supposed expanding universe. Others have presented an explanation that of the red shift is due to the expansion of the universe and space-time, while the red shift (and supposed Doppler effect) is used as proof of the expanding universe. This is a prime example of wrong and illogical circular reasoning.

Chapter 2

Two Initial Fundamental Errors In The Misunderstood Universe

The first error briefly, is due to the assumption that Newton's law of gravity is a universal law of gravity, and therefore is also valid outside our solar system

The second error, with many more serious consequences is the belief that the red shift of Hubble is due to the Doppler effect and that the stars are receding, and also that the red shift can accurately determine distances.

In considering the mysteries of the universe over the last ten years, I became concerned about the lack of true progress in understanding the mysteries of the universe, some going back over seven decades. There is an apparent lack of understanding as shown in the information that is presented by experts in the many recent presentations and publications and awards.

While reviewing in my spare time (perhaps obsessively) much of the available information for the past 10 years, while busy with other scientific and business matters, I gained an insight into the true situation. It is presented here in order to help redirect future work of others, probably to more serious matters such as solution of the energy problem, and for better, cost effective solutions for health problems. Preparing and sharing my insights in this document will permit me to clear my mind of this problem, and to continue to spend more time on my other practical projects.

Because science is said to proceed funeral by funeral, the insight presented here is primarily for future generations of scientists.

I am an applied physicist by inclination and training, and believe that the current theoretical understanding of the universe is based upon wrong interpretations of the observations. Theory is valuable in assembling the observations into a form that can be verified by predictions and then confirmed by observations. If there is a significant difference between verified observations, preferably for observations also verified by others, then the theory must be reexamined and modified.

In around 1514, Copernicus saw and shared some of the truth, and later in around 1609, Galileo saw and also shared some of the additional truth. Even in around 1687, when Newton made his important contribution to the understanding of gravity, he left some more work to be done for cosmic distances probably because he did not have good data for stars outside our solar system. Even Einstein in about 1905 with his theory of general relativity came even closer to the true universe but was apparently also misled about the apparently expanding universe based upon the wrong interpretation of the meaning of the red shift.

Fortunately there remain additional contributions to the understanding of the universe, probably to be made by future generations.

Two initial and fundamental mistakes and assumptions are responsible of the misunderstood universe, and they involve Dark Matter and Dark Energy.

The first mistake is the assumption and statement that Newton's law of gravity is a universal law, also valid at cosmic distances and valid throughout the universe rather than just a law of gravity valid only at shorter distances as in our solar system.

Newton's law of gravity is only based upon observations of planets in our solar system. When applied to observations of motion of galaxies at larger distances, the law fails to explain these observations. It requires unusually large modifications and unproven assumptions such as massive amounts of invisible matter or Dark Matter to supply the necessary gravity.

The second mistake is related to the true meaning of the red shift. Hubble's law is based upon observations of the red shifts associated with remote galaxies and stars and their distances based upon associated determination of their distances. The observations showed that the red shifts apparently increased linearly as the distances increased. Because the Doppler effect could also cause red shifts (and blue shifts in our Milky Way galaxy) it was believed that the observed red shifts were caused by the Doppler effect and apparently indicated that the remote stars and galaxies were receding, although there were no direct confirming observations of the actual receding velocities.

Both of these assumptions are wrong and provide the two major mysteries in the standard model of the universe, Dark Matter and Dark Energy.

As a result of these mysteries there were other consequent errors, assumptions, and questions in the supposed understanding of the universe. Most involve mostly the red shifts, (which contains contributions from gravitational effects).

Some of these consequent errors and questions are: (a) the belief that the stars are receding, (b) the supposed age of the universe, (c) the supposed distance of Quasars, (d) the supposed high power output of Quasars, (e) the supposed angular transverse velocity of some Quasars computed to show transverse velocities larger than the velocity of light C, (f) the time required for forming cosmic structures compared to the supposed age of the universe, (g) the observed gaps in narrow directional views of red shifts, (h) the belief in the expanding universe, (i) the belief in the big bang, and (j) the need for inflation to explain the observed uniformity of the universe. The age of the universe based upon the Hubble constant and assumed velocity is wrong.

Gravity is the largest force in the universe, and without gravity the universe and stars as we know them would not exist because the nuclear processes in stars start under gravitational compression of matter in the stars. Nuclear forces and electromagnetic forces are also strong but only for smaller distances.

Newton's law of gravity is said to be a Universal law of gravity although it is only based upon extensive observational data for the motion of bodies in our solar system. For observations at cosmic distances outside our solar system, the law fails. The concept of Dark Matter is needed to make the cosmic observations fit Newton's law. This is the case for the observed flat rotation velocity curves of spiral galaxies and the earlier observed motion of groups of galaxies by Fritz Zwicky who was not sufficiently appreciated. When we extend Newton's gravitational constant by adding a term linear in distance, r, then Dark Matter is not needed to explain the observations. Our solution is different from the MOND theory of M. Milgrom that involves acceleration.

One basic assumption in the commonly accepted model of the universe is the assumption that Newton's laws of gravity is a universal law and that the gravitational constant is also valid at cosmic distances.

Newton's law and gravitational constant were shown to be valid but only based upon astronomical observations in our solar system and subsequent laboratory experiments.

There is no observational proof that Newton's laws are also valid for the universe or for distances larger than for our solar system. In fact we find that our simple Theory of Additional Gravity (TAG) as an extension of Newton's original law of gravity for large distances that will fit the observations at cosmic distances and without change, will also fit observations in our solar system.

The common assumption about gravity has led to the usually held belief that Newton's law of gravity, combined with observations of motion of groups of galaxies, and the motion of stars in spiral galaxies, need the presence of massive amounts of missing matter, now called Dark Matter to explain the many observations. This has resulted in a number of more serious errors, which are propagated in a number of publications and references by respected astrophysicists.

There are errors and assumptions in the standard model of the universe – and they should be corrected in order to really understand our universe. It is important that the scientific community take careful consideration of the correction of the assumptions in order that many capable scientists do not waste any more of their precious years of research by following ideas and trails that will be discarded by future generations.

A corrected understanding of additional extended gravity outside our solar system (and also in our solar system) does not invalidate the work of Newton and of

Einstein. Their work was based upon observations in our solar system and is still valid there.

There is considerable interest, need, and effort involved in identifying the Dark Matter that is assumed to exist in massive amounts in the universe.

Starting over seven decades ago, there now are serious errors in the commonly accepted model of the universe. Observations of the rotation of spiral galaxies together with early observations by Zwicky of the motion of groups of galaxies, plus Newton's laws, required massive amounts of missing matter, now mistakenly called Dark Matter, to explain the many observations.

In subsequent chapters I will show how my simple extension of the gravitational constant G, for larger distances, r, consisting of an additional gravitational term linear in distance, A*r, can explain the observations without needing the fruitless search for Dark Matter.

Chapter 3

Dark Matter Explained

In general for scientific work, theories are developed to agree with observations, and the value of a theory increases when it can predict future observations. Newton's laws were based upon observations of the motion of planets. Einstein's general Relativity as an extension of Newton's laws did not replace Newton's laws, and are validated by observations of the unusual motion of Mercury, the effect of gravity on photons, and most important were validated by additional measurements of the displacement of star images during an eclipse of the sun.

Fritz Zwicky first introduced the concept of missing (dark) matter in 1933 when he studied the radial velocities of seven galaxies in the Coma cluster. He determined the radial velocities of individual galaxies and the mean velocity of the cluster. From these he was able to make a crude estimate of the total mass of the cluster. Zwicky concluded that most of the cluster mass is in the form of invisible matter or Dark Matter that can be detected through its gravitational effects. A crude estimate at that time was for a ratio of at least 400 for dark mass to visible mass. This was a remarkable accomplishment at that time in view of the limited equipment available. It identified the serious need for a revision of the theory of the gravity in the galaxy. However, a patch in the theory was provided by the introduction of the concept of massive amounts of Dark Matter.

In 1970 Rubin and Ford studied the optical (Doppler shift) rotation curves of the spiral galaxy M31. The results showed a constant rotational velocity at the outer visible portion of the optical disks. This suggested an explanation using a halo of Dark Matter outside the visible optical extent of the disk, similar to the results of Fritz Zwicky for galaxy clusters.

A rotation curve with a constant velocity, plus Newtonian physics, implies that the halo mass increases linearly with r even outside the range of visible stars.

We introduce our Theory of Additional Gravity (TAG) to extend Newton's law of gravity so that it will agree with observations of events at cosmic distances - as well as in our solar system, without needing Dark Matter.

Further study by us showed that the representation of the long range gravitational constant, Ga, can result from the elementary equation describing the equilibrium between the gravitational force, G*M/r*r towards the central mass M at a distance r, and the centrifugal force due to rotation velocity, v, described by v*v/r.

The resulting equation is

$$M*G = v*v*r \qquad\qquad 3.1$$

and for the observed case of constant velocity, reduces to M*G as a linear function of r. The usual interpretation of the situation is that G is assumed to be independent of distance, and therefore the mass M must provide the necessary linear dependence. The unproven halo of Dark Matter around the spiral galaxy was the result. Note that the supposed Dark Matter had the additional properties of not emitting or reflecting light, and of not eclipsing other light.

We realize that there is no reason, without observation proof, to expect the gravitational constant to be constant outside our solar system. With the observational proof, now we can ascribe the linear dependence of M*G to G and show that there is an additional gravitational addition providing the linear component.

Thus $Ga = Gn + A*r$ where Ga represents the additional gravity, Gn is the usual Newtonian gravity, and A is a coefficient for the linear dependence on r. With this formulation, there is no need for Dark Matter to explain the observations.

In our earlier work (about 5 years earlier) we suggested that the flat rotational velocity portion of spiral galaxies could be explained by an inverse r dependence rather than the inverse square dependence on r. We then proposed that the linear term was the first term of a power series expansion of G.

Later, after we used the concept of the balance between the centrifugal force and the gravitational force, it became apparent that the linear dependence on r could be shown directly from the equation without needing a power series approximation.

Our hypothesis of gravitational drag on gas and dust in the interstellar voids together with the additional long-range attractive gravitational force provides a mechanism for photon energy loss, termed "tired light", proportional to distance and will help validate the early hypotheses of Zwicky.

I do not want my theory of additional gravity (TAG) to be confused with the theory of Mordehai Milgrom who had proposed an alternative (MOND) approach to Dark Matter that modified Newtonian dynamics for large distances. He took a different approach involving the acceleration and the masses and segmented regions.

The red shift measurements by Vera Rubin of the motion of stars in spiral galaxies showed non-Newtonian constant rotation velocity of stars at the outer edges of spiral galaxies.

The important work of Rubin, added to the earlier work of Zwicky, plus the application of Newton's principles had led to the amazing conclusion that over 90 percent of the matter in the universe consists of Dark Matter, or invisible mass. Just invoking the concept of Dark Matter to explain the data of Zwicky, and of Rubin has created a need for many complicated theories explaining Dark Matter and fruitless and expensive searches.

Our hypothesis of additional attractive force, particularly apparent at large distances, can explain the data of Rubin and of Zwicky, and, even more interesting, without requiring Dark Matter.

In addition it apparently can explain additional slight attraction forces within our solar system. The very high precision measurements of the NASA Pioneer 10/11 space probes within our solar system appear to support our hypothesis.

Observations of the NASA Pioneer 10 and 11 probes indicated that they were slowing down faster than predicted by Newton's law and Einstein's general theory of relativity. According to reported analysis, "Some extra tiny force - equivalent to a ten-billionth of the gravity at Earth's surface - must be acting inward on the probes, braking their outward motion." Analysis by John D. Anderson and his team at JPL ruled out a number of possible explanations of this extra force.

Our hypothesis provides a Theory of Additional Gravity (TAG) that is significant at cosmic distances but is also apparent at the smaller distance in our solar system as a very tiny additional gravitational component, where they are too small to influence the motion of planets but can slightly influence space vehicles.

The motion of the planets is not measured with the precision obtained for the NASA probes.

We used the limited available rotation curve data for spiral galaxies to estimate the value of the additional force coefficient, A, and the distance, ro, at which the additional gravitational contribution becomes equal to the Newtonian contribution. The values may be different for different galaxies.

As an unexpected result of the hypothesis we were able to predict that this additional gravitational force provides an additional mechanism for the energy loss of photon energy and the red shift of remote stars, in addition to any red shift due to real motion of the stars.

This led us to expand our investigation to include the meaning of the red shift.

Integrating the additional gravitational force equation provides an additional term in the potential energy change. This indicates that the photons (electromagnetic energy) lose energy logarithmically, which is approximately linear in the initial portion of the logarithmic function. It is apparently linearly with distance for the initial smaller part of the distances to remote galaxies.

A major contribution of our hypothesis is to provide a mechanism and prediction that there may not be a need for Dark Matter.

In other chapters we will describe the effect that the additional gravitational force will also have on the red shift, and on the apparent distance based upon the red shift. The result is that the computed energy output of some Quasars (containing black holes) incorrectly appears to be very large, and that the computed transverse velocity of some Quasars, impossibly appear to be above the velocity of light, C.

The wrong value of the gravitational constant will have an effect on the computed time for the formation of cosmic structures such as webs, strings, walls, and voids, and which upon computer modeling conflicts with the supposed age of the universe.

There also is an augmenting effect on the gravitational lenses described by Einstein. Black holes can influence the gravitational lens effect.

Because the distances in our solar system are very small compared to galactic distances, the additional component of the gravitational constant is small and the description of the effect of gravity on planets can be described by the usual inverse square dependence in our solar system, within the sensitivity and resolution of measurements for planets.

Einstein's General Relativity and gravity are also valid in our solar system and beyond.

Four proofs of General Relativity are: (a) Rapid precession of Mercury's orbit, (b) Bending of light (photon path) passing near sun and influenced by gravity, (c) Gravitational red shift in strong gravitational field, and

(d) Time dilation in gravitational fields – depends upon distance from center of earth.

Also the gravitational lenses predicted by Einstein and others have been observed for galaxies.

Chapter 4

Dark Energy Not Needed

Dark Energy is not needed as an explanation for the apparent acceleration of the supposed expansion of the universe. There is no acceleration. Dark Energy also is not needed to provide the suggested massive amounts of Dark Matter based upon the equivalence of matter and energy.

When the distances of the oldest, remote stars were determined from the observed light intensity received and then compared with the distance determined from the Hubble constant and the observed red shift, it appeared that the older stars were much further than expected from their red shifts.

In the standard model of the universe, this was explained by the apparent acceleration of the recession of remoter older stars. The necessary energy to power the acceleration was supposed to be explained by the existence of Dark Energy.

Interestingly, the supposed existence of Dark Energy was used to prove that it supplied some of the vast missing Dark Matter (governed by the relationship between mass and energy described by Einstein's Special Relativity). Previously we showed that there was no need for Dark Matter.

The error is due to the misunderstanding about the red shift. The red shift at large distances is not primarily due to the Doppler effect but is due to three gravitational interactions that remove energy from the traveling photons.

Distance of remote stars is determined by optical intensity (magnitude) and the calculation based upon the inverse square dependence upon distance of the energy received by the observer.

The wrong distance is determined from the red shift and Hubble constant assuming that the linear Hubble relationship can be extrapolated to very, very far distances.

It can be shown that the red shift cannot be extrapolated indefinitely. Actually the extrapolation of red shift to extreme distances means that the photon energy approaches zero. According to the belief in the continued extrapolation of the Hubble linear relationship this results in the conclusion that eventually the photon energy becomes zero, and for further distances, the extrapolated photon energy would become negative. This obviously is not realistic and the solution to the dilemma is that the dependence of the red shift relationship upon distance should saturate for extreme distances.

In fact, the associated analysis of the effect of gravity in reducing the photon energy suggests that for very large distances the decrease in photon energy should have a component that is a logarithmic function of distance. Thus with this model, the photon energy will never go to zero or to negative values because the linear dependence of the red shift on distance must change because of gravitational effects on photon energy.

Chapter 5

Explaining Problems With Gravity

There is considerable interest, need, and effort involved in identifying the Dark Matter that is assumed to exist in massive amounts in the universe.

Starting over seven decades ago, there now are serious errors in the commonly accepted model of the universe. Observations of the rotation of spiral galaxies together with early observations of the motion of groups of galaxies, plus Newton's laws required massive amounts of missing matter, to explain the many observations.

Fritz Zwicky (Zwicky, 1929) first introduced the concept of massive amounts of missing matter in connection with the observed positions and velocities of groups of galaxies. This was based upon the common belief in the universality of Newton's law of gravity in the cosmos.

Subsequently Vera Rubin used the concept of Dark Matter to explain the observed flat velocity rotation curves of some spiral galaxies (Rubin et. al. 1970, 1985). The observed flat rotational velocity curves of stars at the edges of spiral galaxies as reported by Vera Rubin and others could only be explained by massive amounts of invisible, "dark" matter, implicitly assuming that Newton's law of gravity was also valid at large galactic distances.

Thus one basic assumption in the commonly accepted standard model of the universe is the assumption that Newton's laws of gravity and the gravitational constant are also valid at cosmic distances. However Newton's law and gravitational constant were shown to be valid but only based upon astronomical observations and subsequent laboratory experiments in our solar system.

There is no observational proof that Newton's laws are also valid for the universe or for distances larger than for our solar system. In fact we find it necessary for a simple addition to Newton's original law of gravity for large distances to fit the observations at cosmic distances and also in our solar system.

When we extend Newton's gravitational constant by our Theory of Additional Gravity (TAG), which adds a term linear in distance, r, to Newton's gravitational constant Gn, then Dark Matter is not needed. Our solution is different from the interesting MOND theory of Milgrom that involves acceleration.

The usual assumption about gravity has led to the commonly held belief that Newton's law of gravity, combined with observations of motion of groups of galaxies, and the motion of stars in spiral galaxies need the presence of massive amounts of missing matter, now called Dark Matter to explain the many observations. This has resulted in a number of more serious errors, which are included in publications by respected experts.

These errors and assumptions in the standard model of the universe should be corrected in order to really understand our universe. It is important that the scientific community take careful consideration of the correction of the assumptions in order that many capable scientists do not waste any more of their precious years of research, and funding, by following ideas and trails that will be discarded by future generations.

A corrected understanding of additional gravity outside our solar system does not invalidate the work of Newton and of Einstein. Their work was based upon observations in our solar system and is still valid there.

I will show how my simple addition to the gravitational constant G consisting of an additional term linear in distance, A*r, can explain the flat rotational velocity observations without needing the long, unsuccessful search for Dark Matter.

This additional component of the gravitational constant is only easily observable or needed at distances much, much larger than the dimensions of our solar system. It appears significantly at about the size of spiral galaxies, which is around 5-kilo parsec.

Briefly, the inward gravitational force on a mass, m, will balance the outward force due to the rotation of outer stars in a spiral galaxy. The outward force is m*v*v/r where v is the velocity of a star with mass, m, rotating at a distance, r, from the center of the galaxy mass M. The gravitational inward force on the rotating mass, m, is m*M*G/(r*r) where G is Newton's gravitational constant.

The reason for my proposed solution to the mystery can be explained by several fundamental equations.

Thus:

$$M * m * G / (r * r) = m * v * v / r \qquad 5.1$$

This reduces to:

$$M * G = v * v * r \qquad 5.2$$

And for constant velocities, v, then M*G is a linear function of distance, r,

When the linear dependence is assigned to the mass, M, (because everyone knows (?) that the invisible gravity described by G is a constant) we are left with the belief that the invisible Dark Matter is a liner function of the distance, r.

The search for Dark Matter can be ended if we use the above equations to justify an addition to the already invisible Newton's gravitational constant Gn to include an invisible component that increases linearly with distance r. Thus:

$$Ge = Gn + A * r \qquad\qquad 5.3$$

We now do not need to use a power series expansion of the gravitational constant to provide the linear term. The linear term for the gravitational constant now comes directly from our simple equation.

In the outer edges of spiral galaxies Newton's law predicts that the velocity should decrease inversely as the square root of distance, r, instead of the observed flat velocity rotation curves where the rotation velocity, v, is constant and independent of r.

Because of the observed flat rotational velocity of outer stars in galaxies by Vera Rubin and others, (using differential red shift observations) and if we make the usual assumption that the solar system gravitational constant is also valid at galactic distances, the usual result is a need for the mass to increase linearly with distance at the outer location of these spiral galaxies even if there are no stars visible there.

There are no visible stars corresponding to the supposed extra mass called Dark Matter. This Dark Matter shows its supposed existence by the gravitational effects, but surprisingly is transparent to light, does not eclipse visible stars, nor does not reflect light from nearby stars (like the moon reflecting sun light). Calculations show that a massive amount of Dark Matter is required to explain the observations. All sorts of entities with mass are postulated such as but not limited to Wimps, Machos, and Strings, to meet the need for Dark Matter. Expensive research is currently under way to search for the Dark Matter.

Note that under the standard theory of Dark Matter the concept of a belt or halo of Dark Matter OUTSIDE the galaxy is not consistent because Dark Matter would only be needed in the visible region of flat rotation velocities.

Preliminary analysis of data from spiral galaxies NGC2403 and NGC3198 permitted us to find a spiral Galaxy transition radius Rs of 2.7 Kparsec for the transition from the rising portion and the flat portion.

This plus the known value of Newton's gravitational constant Gn (Gn = 6.672 x 10^(-8) cm*cm*cm/gr/sec*sec) gives a preliminary value for

$$A = Gn/Rs = 2.16\ 10^{\wedge}(-26)\ cm*cm/gr/sec*sec).$$

In other chapters we will show the role of my proposed Theory of Additional Gravity (TAG) in a revised model of the universe, and also the implications affecting the concepts of the red shift, which in turn has implications about the meaning of the age of the universe, the expanding universe, Dark Energy, inflation, the CMB, and the Big Bang.

Remember, the razor of William of Ockham (check this true spelling in Google) suggests that the simpler explanation is preferred.

Chapter 6

Pioneer Anomaly Explained

The anomalous motion of the NASA space probes Pioneer 10 and Pioneer 11 can now be considered in conjunction with the Theory of Additional Gravity, TAG, which will explain and predict the puzzling observations.

Over a long time of measurement and reporting, by radio, of the paths of these probes, analysis of the data showed that the paths indicated the presence of unexpected and unexplained, but very small attraction towards the central sun. This force on both independent probes was about many orders of magnitude smaller than expected from Newton's law and gravitational constant.

The analysis of tracking data by radio transmission from the spacecraft indicated that there was a consistent anomalous small Doppler frequency shift corresponding to a constant acceleration of $(8.74 +/- 1.33) \times 10$ exp -8 cm/s*s (or 10 exp -10 m/s*s when cm is converted to m), which is directed toward the sun.

Because there were multiple observations and measurements reported over many years of probe travel, the possibility of experimental error is believed to be small.

The radio Doppler and ranging data, both provided information for the velocity and distance of the spacecraft.

When the gravitational forces are taken into consideration, a very small but unexplained force remains, corresponding to a central acceleration towards the sun of $(8.74 \pm 1.33) \times 10^{-10}$ m/s^2 and for both spacecraft and on opposite sides of the sun.

When the positions of the spacecraft are calculated one year in advance based on measured velocity and known forces (mostly gravity), they are determined to be about 400 km closer to the sun at the end of the year than the predicted distance.

Data from the Galileo and Ulysses spacecraft show similar effects.

As another possible example of additional gravity, the measured value of acceleration for the Cassini vehicle $(26.7 \pm 1.1) \times 10^{-10}$ m/s^2, is about three times as large as the Pioneer acceleration.

A suggestion had been made that the tiny additional forces were due to gas leakage (but this would require a specific direction away from the sun for the gas leakage for both probes).

My knowledge of the small addition to the laws of gravity alerted me to the relevance of reported observations of both the NASA, Pioneer 10 and Pioneer 11 space probes. It was determined that the probes were slowing down faster than predicted by theory using Newtonian gravity. It was reported, "Some extra tiny force - equivalent to a ten-billionth of the gravity at Earth's surface - must be acting on the probes, slowing their outward motion." John D. Anderson and his team at JPL made refined analysis, which ruled out a number of possible explanations of this extra force. The paper of Anderson et al on this subject can be found on the Internet, with a reference provided in the section on References and Reading Material.

Apparently the additional gravity already described by the theory of additional gravity, TAG can help to solve and predict this mystery.

Chapter 7

The Effect Of Gravity On Red Shifts

We will describe the three contributions by the effect of gravity alone in reducing the photon energy and in contributing to the red shift, without the significant need for the Doppler effect.

The three gravitational contributions to the loss of photon energy and the red shift are:

1. Leaving mass against the gravitational force removes energy from the photons. This is seen by the wavelength increase in the case of light radiation from our sun. This is called a gravitational red shift, and is also a factor in global positioning systems, GPS, which require corrections for the gravitational effect on clocks, and electromagnetic, RF, energy.

2. Photons traveling large distances from the source will lose photon energy due to attractive gravity to the source. In the range of distance where the central attractive force is due to the additional gravity, and where the theory of additional gravity, TAG applies with an inverse r dependence of the force, the change of photon energy takes a different form. When the additional inverse distance dependence on r is integrated over distance, the loss of photon energy due to this gravitational component is a logarithmic function of distance and saturates

for very large distances. This is important in the case of apparent Dark Energy associated with observations of galaxies at very large distances.

3. Gravitational drag on the traveling photons has the most effect on the standard model of the universe because it is the reason for "tired light" where the photon energy loss increases linearly with distance, without needing defocusing collisions or the absorption and reemission of photons.

Details related to explaining tired light will be given in later chapters.

Chapter 8

The Effect Of Gravity On The Apparent Quasar Energy

Quasars were discovered in the late 1950s using radio telescopes, and were originally named "quasi-stellar radio sources" and apparently Hong-Yee Chiu introduced the shorter name 'quasar' in 1964.

The unique property of the huge deduced luminosity, together with the apparently large distance demonstrated by the large red shift resulted in some conclusions that need reexamination.

A quasar is a compact collection of stars containing a massive black hole. Because the quasars have a very large red shift, the Hubble law for the red shift results in a determination that the quasars are very distant, and the light received by the observer, when corrected by the inverse square distance reduction of light, suggests that the energy output of the quasar is much greater than that of average galaxies. Some quasars show rapid changes in output suggesting that they are small because of the time required for changes, limited by the velocity of light, to propagate though the quasar.

For the case of Quasars that may have black hole content and provide a massive gravitational force, the effect of the gravity will increase the red shift and will make the Quasar appear to be much further away.

When the apparently further distance is used with the measurement of light emitted by the Quasar, the computed energy output of the Quasar, based upon the inverse square decrease of light, will make the quasar light source and energy output appear to be much larger than when determined for the correct, nearer distance.

This apparently large energy output of Quasars can be incorrect because the gravitational effect of Quasar black holes was not taken into account in determining the Quasar distance.

Chapter 9

Apparently Unusually Large Transverse Velocities For Quasars

Another consequence of the apparent large distance determined from the observed red shift for Quasars with black holes is the unbelievably large calculated transverse (sideways or angular) velocity of the Quasar based upon the supposed distances and the observed transverse angular velocities.

The distance determined from the observed red shift when it is increased by a massive black hole in the Quasar will be larger than the actual distance. When the observed transverse angular velocity is multiplied by the incorrectly determined large distance based upon the erroneous meaning of the observed red shift, the computed transverse physical velocity result is a value greater than C.

According to Einstein, and in conformity with Einstein's insight, there is no possibility (currently) for travel faster than the velocity of light, C.

The neglect of the gravitational effect of black holes, or other massive entity, is responsible for the error in the determination of the red shift, and the distance, and is the cause of the wrong and unacceptable determination of the transverse velocity.

Chapter 10

Formation Of The Cosmic Web – Strings, Walls, Voids

Observation of the arrangements of the galaxies and stars in the universe has shown an interesting web of galaxies with structures such as strings of galaxies, walls of galaxies, and also voids where there are no observable galaxies or stars.

When calculations by others to explain the motion of galaxies to form such structures are reported there apparently was not enough time for such structures to form, based upon the commonly believed age of the universe, about 13.7 billion years, and the use of Newton's universal gravitational constant.

This is an interesting problem for our modern computers and only when the results agree with the observations can this increase confidence in the standard theory of the universe.

Based upon some of our results reported in various chapters we suggest that the age of the universe as based upon the Hubble constant and the observed red shifts need revision because we believe that the red shift for remote galaxies is not based upon the Doppler effect and velocity but only on distance. Also reports on the age of the universe using observations of stars involving radioactive processes indicate that some stars are much older, leading to questions about the age of the universe determined from the inverse of the Hubble constant.

An additional refinement of the computer simulation of the growth of structures in the universe should involve the Theory of the Additional Gravitational constant (TAG) introduced to remove the need for Dark Matter. The additional gravitational force could reduce the time to form the observed structures, and help solve this problem.

In prior postings in my web site, I suggested a mechanism to explain and predict why galaxies would tend to form in strings. Gravitational forces are different from electrical forces because as far as we know, gravitational forces are always attractive in contrast to electrical forces that can be both attractive and repulsive (depending on polarities).

Thus when two galaxies form an initial linear dipole array, their gravitational forces (including the additional linear gravitational component, TAG) will be stronger in the direction of the axis of the dipole. Thus there will be a tendency for additional single galaxies to be attracted along the direction of the stronger dipole gravitational attraction. This will extend the length and the strength of the dipole and increase the ability to continue to attract galaxies in the direction of the augmented gravitational attraction. This is a form of positive feedback that reinforces the process of growing strings and webs of galaxies.

The question of the formation of walls is another problem that may be related to the directional addition of the additional gravitational component of three galaxies in a plane that attracts additional galaxies to the plane. This question will be a future subject for analysis and possibly by others.

The question for the time required for the formation of the observed voids and structure of the universe may be answered when one introduces a longer age of the universe, and also the shorter formation time required because of the additional component of the long-range gravitational force. Massive supercomputers are devoted to modeling the formation of the structure of the universe and their results may be more accurate if the additional gravity, TAG, is used.

(There are many tasks left for future study by theoretical astrophysicists.)

Chapter 11

Einstein Gravitational Lenses - Contributions By The Theory of Additional Gravity, (TAG)

The observations and use of gravitational lenses have confirmed the prediction of Einstein (and possibly Zwicky) about the effect of gravity influencing the path of photons.

These gravitational lenses, when they are closer than cosmic structures will provide magnified images of the structures in the rear. In addition, analysis of the images of sharp rear images can give information about the structure of the gravitational lens itself.

Invisible collections of matter such as black holes can act as gravitational lenses.

The gravitational lenses act through the gravity associated with matter. Thus mapping of the gravitational fields based upon computer analysis of observations can give information about the gravitational fields: are they only the inverse square dependence of the standard Newtonian model, or do they include the longer-range linear component described by our theory of additional gravity (TAG)?

Analysis of the magnified image can possibly give details of the distance dependence of gravity in the gravitational lens: is it just an inverse square gravitational field, or does it include a linear component?

The recent observations of two galaxies colliding (the bullet galaxies) showed the stars stripped away from the two cores that were invisible but showed gravitational lens properties indicating the presence of invisible mass (black holes) providing the invisible source of gravity.

Chapter 12

Gravity Waves Reconsidered

The Laser Interferometer Gravitational-Wave Observatory (LIGO) is supported by the National Science Foundation (NSF), and with an objective for the detection of gravitational waves. The LIGO project is in the relatively new field of gravitational-wave astronomy. LIGO is the largest single enterprise undertaken by NSF, with capital investments of nearly $300 million and having operating costs more than $30 million/year.

Gravitational waves, are supposed to be produced by violent events throughout the universe. LIGO is designed to detect and measure these faintest of signals reaching Earth from space and, at the same time, test fundamental predictions of physics. Its instruments are sensitive enough to measure displacements as small as one-thousandth of the diameter of a proton.

Currently, there are two LIGO facilities, one at Hanford, Wash., and another at Livingston, La. A team from the California Institute of Technology and another team from the Massachusetts Institute of Technology operate LIGO. The collaboration of more than 550 scientists from more than 40 institutions worldwide shows the potential importance. For more information, see: http://www.ligo. caltech.edu/

According to my theory of additional gravity (TAG) the gravitational constant at distances of the size of galaxies will have the usual inverse square component and includes a longer component not in the inverse square format. The usual periodic wave solution in the form of sine and cosine results from an inverse square equation.

In view of the additional contribution to the gravitational constant at distances outside our solar system, the theory of gravity waves should take this into question if gravity waves from other galaxies are to be detected and studied.

Even for gravity waves to be detected originating from our galaxy, the "Milky Way" the linear component of the additional, distance dependent gravitational constant may still be significant.

Will the wave equation be seriously modified by the additional long-range gravity component? The calculation of the expected gravitational signal strength should be refined to take this into effect to see if the sensitivity of the design is sufficient to detect these gravity waves, if they exist.

Chapter 13

Hubble And The Red Shift

The work of Hubble related to the red shift and the determination of the Hubble constant was an important contribution to the early understanding of the universe. Unfortunately the belief by many that the red shift was "apparently due to the Doppler Effect" resulted in many experts going in the wrong direction, and the introduction of wrong and serious errors in the standard model of the universe.

When Vestro Slipher at the Lowell Observatory found longer wavelengths emitted from nebula (spiral galaxies), this could be interpreted as showing that the galaxies were receding, but Slipher postulated that the nebula might only appear to be moving and that the light waves were getting longer as they traveled toward the earth observer.

Georges Lemaitre used the observations of Slipher to suggest that the mass-energy of the universe was initially packed into a primeval atom, and this was the start of the concept of the Big Bang, which apparently agreed with religious beliefs.

A current problem in Astrophysics is related to the meaning of the measurement of distance and velocity of remote galaxies.

There is a question of whether the Hubble constant is a measure of the distance of the stars associated with the observations of red shifts – or is a measure of the receding velocity of the stars. People use the distance interpretation of the meaning of the red shift, and then also used the receding velocity interpretation. Apparently this is related to the belief that the receding velocity is related to the observed distance, and this is not proven.

Actually, there were no direct observations of the receding velocity, but only assumptions of receding velocity based upon the observed red shifts.

We strongly suggest here that the red shifts are related to the distance that the photons travel plus the energy draining effect of gravity, and not at all related to receding velocity. We have shown the important role of the effect of gravity in producing red shifts at large distances, and the relative un-importance of the Doppler effect in the total red shift at cosmic distances. For shorter distances, as in our Milky Way galaxy there are both blue shifts and red shifts, but at larger distances outside our milky way the gravitational component of the red shift dominates.

If our argument is correct, the concept of the receding stars, the expansion of the universe, the age of the universe, and the big bang, which depends upon the velocity interpretation of the Hubble constant and the red shift are probably wrong and must be modified and corrected. Thus the standard model of the universe, believed by most scientists is misunderstood, must be reexamined, and revised. We will contribute to the development of the corrected model.

The background of the Hubble constant and its interpretation is interesting. It appears that there are two different Hubble constants. One is the usual distance Hubble constant determined as a function of observed distances. The other, and wrong, Hubble constant is based upon the assumed contribution of the Doppler effect.

In the 1930s Hubble measured the red shift of remote stars and galaxies along with their distances measured by various means including parallax. Hubble measured the distances of nearby galaxies and their red shifts and showed that the red shifts were proportional to the near distances over the range of his measurements. Subsequent measurements showed that the initial Hubble constant was too large by about a factor of five. Our analysis, now in progress and to be provided elsewhere when completed will show why the Hubble constant for closer galaxies will be greater than for further galaxies.

When examining the history of the use of the red shift for very remote stars as a way of measuring receding velocity, we found that there was another unproven ASSUMPTION and was made by Hubble and others. We learned that the original papers (Hubble and Humason, 1931) had a footnote that indicated that it is not certain that the large red shifts should be interpreted as a Doppler effect but for convenience can be interpreted in terms of velocity and referred to as "apparent velocities." This assumption was later incorrectly converted into evidence of actual velocity and led to serious beliefs about the rapidly expanding universe and subsequently the acceleration of the expansion.

It apparently even misled Einstein who believed (wished) for a flat, non-expanding or non-collapsing universe (open or closed). He then abandoned his cosmological constant - based upon his belief in the work of Hubble that apparently showed an expanding universe.

For very remote galaxies, measurements of the light intensity of Type 1a Supernova are used to determine the distances.

An available plot of red shift as a function of Type Ia Supernova distances was very linear based upon galaxy magnitudes as far out as could be measured, except that there are differences for very far galaxies.

Originally Hubble was concerned with measured distances and red shifts but later it was apparently assumed that the red shifts were caused by velocity and thus Hubble's law became a relationship between distance and velocity, although there were no direct measurements of velocity to correlate to red shift at large distances.

Apparently it is not possible at present to directly measure velocity, (except for angular velocities), for very remote galaxies although distances can be measured. This is a case of theory without confirmation by observation – and in this case is wrong.

Hubble's law and the observed red shifts (and presumably velocities) of remote galaxies showed that the universe was apparently expanding. And even more surprising it indicated that the velocity and expansion were increasing and accelerating at larger distances – because it was assumed that the increasing red shifts corresponded to increasing velocity.

The Hubble constant was introduced to describe the linear dependence of the red shift to the distance, and was then used to calculate the distance to the far galaxies. Note that at no time was the velocity of remote galaxies directly measured but were calculated based upon an assumption that related the velocity to distance by means of a Hubble constant assumed to be valid for large distances.

As will be shown in other chapters, there are three contributions to the red shift that are gravity dependent and that are not velocity dependent.

This supports our opinion that it is not proper to use the red shift to measure the velocity, when only based on the supposed Doppler effect.

Also because the red shift contains three contributions (to be described in other chapters) that are not related to velocity, the use of the measured red shifts to describe the speed of the expansion will give values of the speeds of the expansion that are wrong.

Having opinions and wishes about a static universe or an open or closed universe can lead to assumptions and fudge factors that are not validated by observations. Observations, when interpreted in the wrong way (like the red shift) can be misleading. The Hubble assumption relating red shift to the receding velocity of stars and galaxies is wrong, and has resulted in many wrong assumptions and beliefs.

The three velocity independent contributions to the red shift will question the use of the red shift to measure the velocity according to only the Doppler effect.

Also because the red shift contains three contributions that are not related to velocity, the use of the measured red shifts to describe the speed of the expansion will give values of the speeds of the supposed expansion that are larger than any actual values.

Having opinions and wishes about a static universe or an open or closed universe can lead to assumptions and fudge factors that are not validated by observations. Observations, when interpreted in the wrong way (like the red shift) can be misleading.

Edwin Hubble showed in 1929 that the more distant the galaxy, the larger this "red shift." Astronomers traditionally have interpreted the red shift as a Doppler shift and deduced that because the galaxies apparently recede from us this indicates an expanding universe. Because the red shift is usually incorrectly expressed in terms of velocity, this error about the expanding universe continues to date.

Initial studies and reported observations suggested that galaxy red shifts take on preferred or "quantized" values. These discoveries led to the suspicion that a galaxy's red shift may not be related to its Hubble velocity alone. I see no reason why if the red shift is due to receding velocity that the velocities should be quantized, but can understand that it could be quantized if it depends upon distance. Observations show that there are gaps in the distribution of stars and galaxies.

For reported observations of several well-studied galaxies, including M51 and NGC 2903, these observations showed two distinct patterns of red shifts. The observations showed that the jump in red shift between the spiral arms always tended to be around 72 kilometers per second, no matter which galaxy was considered. Subsequent observations and studies apparently showed that velocity breaks could also occur at intervals that were 1/2, 1/3, or 1/6 of the original 72 km per second value.

The 72 km per second discontinuity reported by others should have been followed up. The accuracy of the data at that time was not enough to show the effect clearly. Also, there was no reason to expect such an effect.

Not surprisingly, it was reported that the first papers from the observers presenting this evidence to the astronomical community was not accepted.

For various observations, the observed differences in red shift between pairs of galaxies tend to be quantized rather than continuously distributed. The red shift differences are observed to be in near multiples of 72 km per second. The observations were apparently not sufficiently accurate to be definite.

Steven Peterson, using telescopes at the National Radio Astronomy Observatory and Arecibo, published a radio survey of binary galaxies made in the 21-cm emission of neutral hydrogen, which can be made more accurately for the 21-cm line than for lines in the visible portion of the spectrum. Specifically, red shifts at 21 cm could be measured to accuracy better than the 20 km per second, and this permitted better detection of a 72 km per second periodicity.

Radio astronomers have examined groups of galaxies as well as pairs. There is no reason why the quantization should not apply to larger collections of galaxies, so red shift differentials within small groups were collected and analyzed. Again a strongly periodic pattern was observed by others, confirmed, and reported.

The tests described were for narrow angle directed views of the sky, and showed periodicities or gaps in the red shifts. It is predicted by us that when the other narrow angle views were overlaid to average the red shifts over a wider view of the sky, the periodicities and gaps will disappear. This is similar to the case where when a cross section of a sponge will show random gaps or voids, the voids will be averaged out when the cross sections are overlaid.

Current cosmological models cannot explain this grouping of galaxy red shifts around discrete values in the universe. The discrepancies from the conventional picture will require that the entire beliefs in the standard model of the universe and of cosmology would be reconsidered.

Several ways can be conceived to explain this quantization. As noted earlier, a galaxy's red shift is not due to a Doppler shift, that is the currently commonly accepted interpretation of the red shift, but there can be and are other interpretations.

We suggest here that the galaxy's red shifts may be a fundamental property of the structure of the universe. Briefly, the distribution of stars and galaxies are arranged in a web structure containing strings, walls, and voids. If as we theorize that the red shifts are associated with distance the photons travel (rather than because of velocity) there will be gaps in the distances traveled when viewed over narrow directions. Thus there will be gaps in the distribution of distances and the corresponding red shifts.

Tifft provides information about observed gaps in red shifts. (Tifft, 1987)

When one views the spatial distribution of galaxies in the universe, it is apparent that there are separations between the galaxies and in the web of stars.

This suggests that the gaps in the distribution of red shifts is due to the gaps in photon travel distance, and this will be seen in a narrow angular view of the star distribution. When looking in different directions, the locations of the stars will be different from other directions, and it is predicted that the gaps will be smoothed out when different views are overlaid for a larger view of the universe and when the total array of gaps are evaluated. Others, however, will erroneously present this as proof that the gaps do not exist.

Chapter 14

Red Shift Components Involving Gravity

Many of the errors in the standard model of the universe are related to not understanding that the red shift for large distances is largely due to the effect of gravity in draining energy from traveling photons. In general the red shift includes a small contribution due to the Doppler effect, and many wrongly believe that the Doppler effect is the real reason for the total red shift. For closer distances such as in our Milky Way there are both blue shifts and red shifts, which can be seen because the distance component of the red shifts is smaller.

However, the red shift is actually due to three ways that photons moving against gravitational forces will lose energy, and have their frequency decrease, with a corresponding increase in wave length, while the velocity remains at C, the velocity of light.

The first way that gravity contributes to the red shift is when photons leave the sun; they experience the well-known gravitational red shift.

The second way involves our Theory of Additional Gravity (TAG) that adds a linear component to Newton's gravitational constant. Thus when photons travel long distances against gravitational fields; the work done as determined by integrating the gravitational force (including the additional component A/r already described) determines the red shift. When the gravitational force, $M/r*r + M*A/r$ is integrated with respect to distance r, this results in a term $M/r + M*\ln r$. The first term decreases quickly as distance r increases, but the second term is a much slower, increasing logarithmic component.

An interesting and important consequence that shows a need to modify Hubble's work on the linear red shift dependence is that normally the linear dependence of the red shift when extrapolated to the limiting distance case of very large distances would eventually increase to the point where the decreasing photon energy would become zero and then even negative - which is very unlikely. With the logarithmic dependence on distance in our model of the universe, the linear plot of red shift and photon energy vs. distance for very far distances would curve away from linear and thus not need become zero.

It should be noted that the analysis of this component of the contribution of gravity the red shift explains and predicts the apparent acceleration of the supposed receding of very remote galaxies, and the supposed need for Dark Energy..

This supposed accelerating recession is due to the belief that the distance of the very remote galaxies as determined from their observed reduced optical magnitude is much greater than the distance determined from their observed red shift, and that apparently these remote galaxies experience acceleration. The energy needed to power the supposed acceleration was termed Dark Energy and was the second major mystery of the universe.

Showing the consequences of initial errors, others have used the Einstein relationship between mass and energy ($E=m*c*c$) to suggest that the supposed Dark Energy can supply some of the missing mass of Dark Matter.

In my opinion, invoking a mystery is the same as saying "I don't know, and may be wrong."

Our solution of the Theory of Additional Gravity, TAG, for the first mystery, Dark Matter, which adds a linear term A*r to Newton's gravitational constant G apparently has contributed to the solution of the second mystery, Dark Energy.

As an aside comment, we point out that others have suggested that the supposed amount of Dark Energy can be related to Dark Matter according to Einstein's relationship between mass and energy and thus have compounded the errors..

The third way that gravity contributes to the red shift, in my opinion, is the most important because it provides the prediction and explanation of tired light as proposed by Zwicky, and which many others ignored.

It is observed that the red shift is proportional to the distance of travel of the photons, (at least for distances not near the logarithmic range, as described above). The red shift is presented as evidence that the galaxies are receding, leading to the belief in the expanding universe, caused by the supposed big bang.

I have seen explanations that the increase in photon wavelength proving the expanding universe is caused by the expanding universe. This is an example of wrong circular reasoning.

The red shift is supposed to be due to the Doppler effect showing that the galaxies are receding, although there are no direct observations of the velocity. I will provide an alternate mechanism involving gravity that will predict and explain how gravity can result in the linear dependence of the red shift on distance. As a bonus this will explain, in other chapters, the cosmic microwave background CMB and the low temperature of the electromagnetic radiation.

As photons travel through interstellar space and pass near the vast amount of gas and dust in interstellar space, there is gravitational interaction between the mass of the photons and the small masses of gas and dust. (Note that photons have mass determined by their energy according to Einstein's special relativity, which has been proven explosively.)

The gravitational interaction for transfer of energy is stronger when the masses are close in value. The event horizon limits the gravitational interaction of very remote masses, in spite of the much longer range of the linear additional gravitational component and this eliminates a mathematical problem when the total number of gas and dust entities outside the event horizon would be gravitationally involved.

The cross section for physical contact with the gas and dust is much smaller than the cross section of the gravitational interaction. (The calculation of the relative cross sections is left as a task for degree candidates.) The transfer of kinetic energy from the photon to another mass is much more effective when the masses are nearly equal. As a photon travels and approaches near the gas and dust material the gravitational attraction will tug the gas or dust and will impart kinetic energy to them (without needing collision or absorption) and the photon will lose some of its energy thus contributing to its red shift.

Others have proposed that collisions of photons with gas and dust can involve absorption and reemission of photon energy as an explanation for the observed red shifts. We suggest that this in not the case because, as others have objected to this explanation, the resulting images would be fuzzy, and however the remote images provided by the photons are observed to be very sharp.

This transfer of energy by gravitational forces without collisions is similar to the case of energy transfer by our moon to our earth through gravitational tides. As a result of the transfer without actual contact, the tides gain energy, and the moon loses kinetic energy, slows down, and slowly increases the diameter of its orbit around the earth.

For photons passing close enough to the gas and dust in interstellar space the gravitational interaction will be stronger and they will come into thermal equilibrium. Other photons not as close will pass without interaction and will continue to the radio frequency range with longer wavelengths.

Those photons that interact stronger with the relatively slower gas and dust will be thermalized to the temperature of about 2.7 K degrees of the gas and dust. This interstellar temperature was previously estimated to be about 5 K degrees by others as based upon the light energy available from the visible stars. (See the references and reading material, particularly Assis, A.K.T., History of 2.7 K) Do a Google search on "History of 2.7 k Temperature Prior to Penzias and Wilson".

The temperature of interstellar space was calculated earlier from the energy in the light from stars. Early workers on the low temperature of space included Eddington, Guillaume, Regener, Gammow, and Nernst, who did not use or need the concept of the big bang.

Photons at a temperature of 2.7 K will provide the observed CMB, and photons supposed to be those cooled from the big bang are not needed as an explanation and are not proof.

The gravitational interactions between the photons and interstellar material should make the photons accumulate with a higher density at the microwave frequencies related to the 2.7 K temperature and will be greater than for the infra-red and radio frequency photons, as is observed.

We now have an explanation for the observed red shifts being a linear function of distance (tired light) along with a bonus explanation for the CMB and the low temperature for the microwave radiation.

Thus, the understanding of the role of gravity has provided better understanding of Dark Matter, Dark Energy, the red shift and tired light, the CMB, the 2.7 K temperature, apparent expansion of the Universe, and the big bang.

We believe that the above explanations are based upon basic physical principals together with observations reported by others, and are not pure speculation. It is able to clarify and replace many current beliefs which themselves are pure speculation, unsupported by observations.

Chapter 15

Tired Light, Red Shift, And Gravity

When Hubble, measured the red shifts of remote galaxies and stars along with their distances determined from other observations, it was believed that receding galaxies caused the red shifts. (Hubble, 1929, 1936, 1937).

Actually, Hubble only measured distances but did not measure velocities; it was assumed that the Doppler effect caused the red shifts and thus showed that the galaxies were receding and that the universe was expanding.

Fritz Zwicky (who was not sufficiently appreciated) proposed an alternate explanation. (Zwicky, 1929) This involved his concept of tired light where photon energy was lost and red shifted in traveling large distances.

Attempts by others to explain tired light by photon collisions with entities in interstellar space were not successful because they involved scattering or absorption and re emission that would produce blurred images rather than the sharp images associated with the observed red shifts.

The introduction of failed explanations should not remove the validity of other explanations that are more valid.

We will present our explanation for the red shifts that uses the three contributions of gravity alone in producing red shifts, in addition to any minor Doppler effects.

1. Red shift due to loss of photon energy against the gravitational force of the sun or of a large mass

of a galaxy or quasar or black hole in a galaxy. This is well known and accepted.

2. Traveling large distances against gravity from the photon source to reach the observer.

3. Photons traveling large distances to the observer and losing energy through gravitational drag of gas and dust in interstellar space without blurring collisions.

When traveling large distances, the photons will move against the additional gravitational force described in our Theory of Additional Gravity (TAG). This theory provides a long-range gravitational force depending inversely on distance (in addition to the shorter inverse square distance dependence). When this inverse distance force is integrated over distance to determine the energy lost, the result is that this contribution to the red shift caused by this energy loss is a logarithmic dependence on distance.

It can explain why a linear extrapolation of the linear Hubble dependence on distance cannot experience the apparent conflict that would require zero and even negative photon energies when extended to a sufficiently large distance.

It can also explain why there is an apparent acceleration of the supposed receding velocity of very remote galaxies, and the wrong need for Dark Energy. This is discussed in more detail in other chapters.

The most important aspect of tired light as a linear function of travel distance is due to photons traveling large distances and losing energy through gravitational drag of gas and dust in interstellar space. This does not involve actual collisions of photons with gas or dust but involves the gravitational interaction of the fast photons with the slower gas and dust and which adds motion to the gas and dust and thus extracts photon energy resulting in a lager red shift.

To help understand the science involved, consider the case of our moon and the earth. The motion of the moon caused tides on earth, which extracts energy from the motion of the moon without actual collision. The result is that the kinetic energy of the moon decreases, and the distance to the moon increases slowly.

The extraction of photon energy through gravitational effects is similar.

In another chapter we will use this process to provide an alternate explanation of the Cosmic Microwave Background, (CMB) and the achievement of the 2.7 K temperature of the CMB.

Chapter 16

Consequences Of The Misunderstood Red Shift

The misunderstanding of the meaning of the red shift as measured by Hubble and others is more serious than that associated with Dark Matter.

In addition to the introduction of the mystery of Dark Energy, the red shift error caused much more serious errors, confusion, and misunderstandings.

I believe that as a result, over the recent years the scientists and engineers were led into areas and searches that will be much less productive.

The following is provided to help those less acquainted with astrophysics.

Light emitted from stars will have characteristic emission or absorption spectral lines associated with elements in the stars. When examined by a spectroscope that permits examination of the spectrum, it is sometimes noticed that the wavelengths of the spectrum for light from a star are shifted towards longer wavelengths compared to that from wavelengths seen in laboratories. This is called the red shift and corresponds to lower photon energy.

(Note that there remains a problem in understanding the observed different time decay curves of supernovae that can range from 20 to 40 minutes.)

Determination of the distances of galaxies including distances obtained by observations of their magnitude (light intensity received at the observer), together with observations of the associated red shift provided a linear relationship between the red shifts and the deduced distances.

The Hubble constant describes this relationship, and is equal to the ratio of the red shift divided by the distance. The problem with the Hubble constant is due to the wrong and common belief that the red shift is due to the well-known Doppler effect (where velocity can increase or decrease the frequency and wavelength of signals).

Hubble initially described the red shift "an apparent Doppler shift" although he later indicated reservations about this description. Actually the velocities were never directly observed and only distances were measured

This wrong association of the red shift with velocity has serous consequences for the correct understanding of the universe. One consequence was the belief that the red shift showed that the stars and galaxies were receding, and that the universe is expanding.

This fooled Einstein who believed (hoped) that the universe was flat and who introduced his cosmological constant into his equations to provide a fiat universe solution. He was reported to later call this his greatest blunder.

Because the Hubble constant is not associated with velocity, the age of the universe deduced from the reciprocal of the Hubble constant is wrong.

Following the belief that the stars and galaxies are receding and that the universe is expanding (open universe), the current big bang concept logically followed, but was based upon a wrong understanding of the meaning of the red shift.

Also my personal opinion is that a theory, when it is extrapolated to limits and then results in a singularity (like the start of the supposed big bang) this singularity indicates a conflict in the theory and that the theory must be reexamined and modified to remove the singularity.

Actually in addition to the inability of the red shift to support theories related to supposed velocity, it also has limitations in its ability to describe some observations associated with distances. Reported observations of gaps in the distribution of red shifts (by Arp and others) in narrow directions were rejected, probably because they are not in agreement with the standard theory of the red shifts.

Also, the observations (by Arp) of galaxies that apparently are in close proximity because they were visibly connected by streams of stars, were questioned because these galaxies also had significantly different red shifts apparently showing a large distance in separation.

Actually this could be understood if it is recognized that the presence of a large mass (like a black hole) in the apparently further galaxy could contribute and make the red shift larger.

The most serous consequence associated with incomplete understanding of the red shift is the apparent need for the mystery of Dark Energy.

As an aside, my opinion is that if one calls some idea a mystery it is a gentle way of saying, "I don't understand this and it is probably wrong but I still stick to the idea."

The idea for Dark Energy is the fact that for very remote (receding?) galaxies the distances determined from star magnitude is larger than expected from the distance deduced from the associated red shift and apparently indicated acceleration of these remote galaxies.

It was then suggested by experts that because the speed of the further receding galaxies had apparently increased

Dark Energy was needed to provide the additional energy required.

As a further insult to the understanding of the universe, it was suggested that because of the Einstein relationship between energy and mass, this supposed Dark Energy could supply much of the missing Dark Matter. However, we had already shown that Dark Matter is not needed when we introduce the Theory of Additional Gravity (TAG) in Newton's law of gravity.

Actually the difference between the distances of remote galaxies as determined from the light received (magnitudes) and that determined from the red shifts is because the linear Hubble relationship cannot be extended or used very large distances. A simple argument will demonstrate this.

As the distance and red shift increases, the photon energy decreases, and according to extrapolation of the common linear belief of the Hubble relationship the photon energy will eventually become zero and then negative. This obviously is not reasonable and suggests that the red shift dependence on distance must become non linear, or be reexamined.

As an example, in our earlier analysis of gravity, we showed that our Theory of Additional Gravity, TAG, suggests that for large distances the additional energy loss (red shift) shows as a logarithmic function of distance.

Without realization of the limitation of the shift at very remote galaxies, the evidence for the accelerated velocity of the older remote stars is not valid, and the need for Dark Energy does not exist, and the mystery of Dark Energy disappears.

If the reader believes the arguments presented here then the corrections of other aspects of the misunderstood universe has begun, and could result in a truer understanding and the removal of some mysteries.

The resulting errors in the standard model of the universe are discussed in following chapters. We leave some additional revision of the standard model of the universe to future generations, and possibly for the current generation.

Chapter 17

Galaxies And Stars Are Not Receding

In contrast to the current belief in the standard model of the universe where it is suggested that the observed red shifts are evidence that the galaxies and stars at large distances are receding from the observer, we have provided an analysis of the components of the red shift to explain that the red shift does not support the supposed motion.

Our analysis showed the reasons why the red shifts are not mostly due to the Doppler effect for large distances and are not due to velocity.

Thus the belief in the receding galaxies and stars should be abandoned, or revised, as well as for belief in the big bang.

This is very simple statement but it is important for others to understand the implications for future working the field of cosmology.

Details for this conclusion are provided in other chapters.

Chapter 18

The Universe Is Not Expanding

Because our analysis of the true meaning of the red shift shows that the remote galaxies and stars are not receding, it also shows that there may not be proof that the universe is expanding.

If the universe is not expanding then the belief in the big bang must be questioned.

Also the concept of inflation is a key idea in cosmology but according to the chapter on inflation, another argument is presented to explain the observed uniformity of the universe without the need for inflation in a small fraction of a second – which cannot be verified because it occurred long ago in the past, requiring and enabling pure speculation.

The scientific community should be made aware in this situation, and admit it, so that attention can be paid to more pressing serious scientific matters.

More detailed discussions of the expansion and the uniformity of the universe are provided in other chapters.

Chapter 19

Flat, Open, Or Closed Universe

There is a question about if the universe is flat (not expanding or contracting), or open (expanding), or closed (contracting).

Einstein wanted to believe that the universe was constant and introduced a stabilizing cosmological constant into his relatively equations.

Alex Friedman, in 1923, published his solution that showed an expanding universe model based on Einstein's 1915 General Relativity theory. He simplified an equation and solved it and got the expansion and this provided support for the big bang. But there are no actual observations that prove the expansion or big bang. Simplification can ignore potentially important factors.

Later, George LeMaître also supported the concept of the expanding distances, and the big bang.

Einstein clung to the static model of the flat, non-expanding universe, but changed his mind after the work of Hubble and the Doppler effect interpretation of the red shift, which apparently showed that the universe is expanding.

Actually there is no practical reason for knowing the eventual end of the universe many billion of years in the future, except for curiosity. The answer will not matter to humanity.

There, however there is one possible reason to verify the fate of the universe and that is to provide some confidence that other beliefs about the university are consistent and probably correct.

However, there do not appear to be any possible observations that could be used to test the beliefs, speculations, and theories about the eventual future fate of the universe.

Remember, equations, beliefs, or wishes are not proofs.

Chapter 20

Supposed Expansion Of The Universe

As we discussed in other chapters in this presentation, the incorrect assignment of the Doppler effect as the cause of the red shift has led to a number of serious misunderstandings.

By now we should agree that the concept of tired light as a linear function of distance can be explained as a consequence of the effect of gravitational drag by interstellar gas and dust on the photons traveling nearby, to reduce photon energy, (increasing red shift) without needing collisions.

Once we understand this, it is obvious that the observed red shifts cannot be used to demonstrate that the galaxies are receding.

Because the galaxies are not receding then the conclusion must be that there is no observational proof that the universe is expanding.

Interestingly without proof that that the universe is expanding, there is no need for the big bang.

Many people have claimed that there was no big bang but apparently without convincing reasons.

The big bang concept is a firm part of the standard model of the universe; along with other concepts we have discussed and (perhaps) shown to be false, and needing reexamination.

The next generations will have to consider the true meaning of the question about the big bang.

There is no need for a rush to decision because the answer has no effect on present science, except for the possibility that the current generation of scientists and engineers could have used their funding, training, skills, and effort on more significant and present problems.

The funds are limited and could be used for solving the current problems such as (a) energy replacement, (b) reduction of health care costs, (c) global shortage of water, (d) global shortage of food, (e) global temperature changes, (f) damage by storms, and many more that will turn up.

However, there is a residual and indirect benefit to work on the standard model of the universe, because the resulting scientific and engineering progress, training, and experience frequently can be used to solve other problems.

Chapter 21

Big Bang Is Not Proven

The belief in the big bang is an important part of the standard model of the universe, and it may be difficult or impossible to correct this belief, at least until the next generations of scientists. We shall try to add our contribution here.

There apparently are many who have added their name to an Internet document questioning the reality of big bang, but scientific proof is needed to support their concern. We hope that our analysis of the misunderstood universe will provide support for their concern.

Many indications of the existence of the big bang have been cited including the (a) cosmic microwave background, (b) the 2.7 K temperature of the microwave radiation, and (c) the concentrations of hydrogen, helium, and lithium in the universe.

The Big Bang theory developed from observations of the structure of the universe and from theoretical considerations.

Friedmann, a mathematician, derived the Friedmann equations from Albert Einstein's equations of general relativity. He showed that the universe might be expanding or collapsing. Einstein had added a cosmological constant to his equations to support his belief (wish?) for a static universe.

Vesto Slipher, in 1912, measured the first Doppler shift of a "spiral nebula" (spiral galaxies), and found that many nebulae were apparently receding from Earth.

Edwin Hubble, starting in 1924, measured the distances to the spiral nebula and the associated red shifts and showed a linear relationship and introduced the Hubble constant to describe the linearity.

Georges Lemaître, in 1927, predicted that the recession of the nebulae (galaxies) was due to the expansion of the universe, and suggested that the universe had started from a single point and expanded in a small fraction of a second (the big bang). In general, I believe and suggest and also speculate [sic] that theories that include singularities are suspect and based upon speculation.

Hubble, in 1929, demonstrated a linear correlation between observed distance and the red shift (believed to be an indicator of recession velocity). This provided support for the work of Lemaître and the expansion of the universe, and the big bang, which are now firmly imbedded in the beliefs of the establishment.

During the 1930s other ideas were proposed to explain Hubble's observations. This includes the important tired light hypothesis of Fritz Zwicky, (which I support and can explain).

Fred Hoyle's proposed a steady state model, where new matter would be created, in the voids created as the universe seemed to expand. In this model, the universe is roughly the same at any point in time.

Lemaître's Big Bang theory was supported and developed by George Gamov, who introduced big bang nucleosynthesis. His associates, Ralph Alpher and Robert Herman, predicted the cosmic microwave background radiation (CMB).

Hoyle derisively introduced the name for the big bang as "this big bang idea."

The discovery and confirmation of the cosmic microwave background radiation, CMB, in 1964 apparently supported the Big Bang theory for the origin and evolution of the cosmos. Much of the current work in cosmology now is now based upon this belief.

Additional data about the universe is obtained from advanced telescopes as well as the analysis of data from satellites such as COBE, the Hubble Space Telescope and WMAP.

Cosmologists recently (compared to long ago history) have made the unexpected discovery that the expansion of the universe appears to be accelerating, requiring the concept of the mysterious Dark Energy to power the acceleration.

In our analysis and presentations, we have shown that the Hubble measurements are only related to distance and not to velocity in spite of the concept of the "apparent Doppler effect." Thus there are no direct observations of the recession of galaxies, and no support for the expansion of the universe. Without expansion being proven, there is no justification for extrapolation back in time to a singularity followed by a big bang.

In other chapters we will predict and explain the cosmic microwave background and the 2.7 K degree temperature. As part of my experience I have designed, built, and used a number of different microwave systems for published research, and one of my minors in physics at MIT was electromagnetic theory.

The concept of Dark Matter was clarified in other chapters to show that the linear red shift dependence on distance could not be extended indefinitely; otherwise it would eventually require zero or negative photon energy.

There must be a deviation from linearity at large distances and the apparent distance determined from the red shift will be less than that determined from light intensity. This will explain the apparent and incorrect acceleration of the supposed recession of remote galaxies.

There are two errors in the belief of acceleration of the expansion: first is the error in distance determination from the red shift, and second is the wrong assumption that the red shift demonstrates recession of galaxies.

Many of the ideas presented in different chapters include repetitions of material presented in different ways in other chapters. This is done to insure that the ideas are recognized, and to permit chapters to be absorbed independently of the other chapters, although other chapters can be read to provide additional details.

Chapter 22

Age Of Universe Not Correct

The age of the universe currently is believed to be 13.7 billion years. It is based upon the Hubble constant relating the observed red shift to the observed distance of remote galaxies.

The problem and error is due to the belief that the red shift is due to the Doppler effect and that it shows the receding velocity of the galaxies.

The age of the universe is calculated from the reciprocal of the Hubble constant (velocity/distance) to give a number with the dimension of time.

Unfortunately, the concept of velocity in the Hubble constant is not supported by observations of velocity, but just the assumption of a Doppler effect.

Other independent determinations of the age of older stars show traces of nuclear process that indicate the possibility that these stars are older than the supposed age of the universe.

The question of the age of the universe is still an open question.

However, the age of the universe is necessary in knowing the event horizon. The event horizon is the distance that effects and events can travel, with an upper limit of speed of C (the velocity of light), during the life of the universe.

The event horizon is an important factor in the consideration of the theory of inflation that is supposed to explain the uniformity of the universe. The need for the inflation theory is related to the ability of portions of the universe further than the event horizon to interact and come into equilibrium with other portions of the universe further away than the event horizon. The size of the universe is much greater than the event horizon corresponding to the accepted age of the universe.

This concept of inflation will be discussed and explained in more detail in other chapters.

Chapter 23

Gaps In Narrow Views of The Red Shifts

As an example of the resistance of the experts in the field of astrophysics in rejecting new insight related to the standard model of the universe, I read of some interesting observations by another scientist of gaps in red shift distributions that apparently were rejected by a professional publication. However, other observers also reported these periodic gaps.

In one of the books by H. Arp, (Arp, 1987, 1998) he recounted the observation of periodic gaps in the observed red shifts, apparently confirmed by other observers. As described in the book, when he submitted a paper reporting the periodic gaps, the editor refused to accept the paper apparently replying in effect that the periodic gaps were not observed when the astronomical telescopes viewed the whole sky.

However, if one views the distribution of red shifts along a narrow angle direction and then views the distribution along a different direction, it is expected that the distribution of gaps will change. As a result it is expected that when overlays of multiple directional scans are used this will give an average where the periodic gaps will be filled and the gaps will not be observed.

With respect to the gaps in the red shifts, and if the red shifts are related to gaps in receding velocity there is no reason presented to explain the velocity gaps.

However if the red shifts are related to distance traveled, as described in other chapters, then it is expected that there should be gaps in the red shifts due to gaps in the spacing and distribution of galaxies. The distribution of gaps in the array of galaxies will not be the same in all the viewing directions, and the distribution of photon travel distances and red shift gaps seen by the observer will not be the same in all directions.

The reported periodic aspect can be explained by a degree of uniformity for distribution of galaxies in the universe.

This should explain the observations but does not explain the reported rejection of Arp's paper by the editor.

Actually, this is not a serious problem because the truth will eventually surface, but the apparently unwarranted actions of the editor could be a barrier to efficient scientific progress.

Chapter 24

Connected Galaxies Having Different Distances And Red Shifts

When two galaxies are observed to be in close proximity as is shown by observation of a stream of stars moving from one galaxy to the other, the two galaxies may have significant differed red shifts showing surprisingly different separations.

In one of the books by H. Arp, (Arp, 1987, 1998) he provides a photograph that shows two galaxies connected by the stream of stars indicating their close proximity. He reports that the observed red shifts for these galaxies are also significantly different - apparently proving a conflicting large separation in spite of the apparent connection.

Others apparently have questioned his data. Identification of his related books is available in the section for Referenced and Reading Material if you want more details.

I suggest that the apparently further galaxy falsely appears further because it contains extra massive matter such as a black hole whose strong gravity adds to the red shift so that the Hubble relationship (incorrectly ascribed as due to the Doppler effect) will indicate a larger distance. This shows a need for better understanding of the features of the universe.

Actually, the red shift is a measure of distance the photons travel, and can be increased due to the photon energy being drained when moving away against the strong gravity produced by a source in the galaxy consisting of a large mass such as a black hole.

These galaxies are really in close proximity in spite of the error in the interpretation of the different red shifts.

This can be another example of the misunderstood universe, and the difficulty of getting the established experts in accepting solutions.

Chapter 25

Inflation Is Not Needed

It has been shown in other chapters that there are no observations showing that the galaxies are receding, or that the universe is expanding. There is a misinterpretation of the meaning of the red shift.

Previously when it was believed that the universe was expanding, and then extrapolating back in time, this suggested that the expansion started from a singularity with a big bang.

Because current observations showed that the universe was uniform in all directions, it was necessary to explain the uniformity in spite of the limitations of the event horizon. The concept of the event horizon is based upon the accepted limitation of the velocity of light, C, and the assumed age of the universe. If the universe is larger than the supposed event horizon (very likely but not proven) then there was not enough time for the structure of the universe in one region, A, further away from another region, B, to equilibrate with region B. This is true whatever the age of the universe and size of the event horizon are.

The observed uniformity of the universe needed to be explained in spite of the fact that there was not enough time, limited by the age of the universe, for conditions in other parts of the universe to reach and equilibrate with the other parts outside its event horizon, in view of the limitation of the speed of light.

In order to resolve this problem, Guth of MIT discussed the interesting concept of Inflation. (Guth, 1997) According to this model, in the beginning of the (supposed) big bang, and in very tiny fractions of a second, the universe was very small, and inflation in this fraction of a second permitted the observed uniformity to occur before the universe had later expanded to the size that would prevent uniformity within the time limit of the age of the universe.

In other chapters I explained that there are no observations of the velocity of the receding galaxies, or of the expanding universe, or of the big bang.

Actually I can provide the following alternate mechanism for uniformity of the universe, with or without needing the big bang concept or the inflation concept.

Assume a region of the universe, A, which itself is uniform because it is smaller than the event horizon (whatever the age of the universe is), and regions B, and C which are similarly uniform.

If the regions are selected so that there is a partial spatial overlap of region A with B, then all parts within A and B are uniform. A similar argument occurs if C partly overlaps with A or B. Thus A, B, and C regions are uniform. This argument can be extended to all regions connected with A, B, and/or C that is needed to cover the total universe.

Because of the mathematical and logical argument: if A=B and B=C, then A=C. This can be extended to the other regions to cover the universe.

Of course if new regions of the universe are continually created (unlikely in my opinion and apparently without observational proof) the analysis must be reconsidered.

In any event, in my opinion and based upon the information presented here and in other chapters, there is no big bang, and no need for the concept of inflation.

Chapter 26

Olbers' Paradox And The Dark Sky

Long ago, Olbers' described a paradox associated with the observed dark sky. If the universe is infinite with an infinite content of stars emitting light energy, why is not the sky completely filled with sources of visible light giving a bright sky rather than a sky mostly dark?

One explanation proposed by others is that for the stars that are further away than the event horizon, their light cannot reach the observer.

The concept of the event horizon arises because of the limit on the velocity of light, C, and the age of the universe; light cannot travel from more than the distance determined by the product of C and the age of the universe.

Actually the age of the universe as determined from the Hubble constant (when assumed to be a measure of velocity) is not correct, as is shown in other chapters.

Another proposed explanation by others is that other suns in the path to the viewer block the light from further stars. However the light from intervening star replaces the blocked light.

The presence of interstellar gas is proposed to explain the loss of light from star sources. The red appearance of many stars is explained by the scattering of light by interstellar gas, similar to the red appearance of the setting sun.

Another suggestion by others is that the light from the remote stars is absorbed by the gas and dust in the interstellar space on the way to the observer. This has been countered by the argument that the blocking gas and dust would equilibrate with the blocked sun and in turn would give a bright background in the view of the sky.

The hypothesis presented here by me can provide an explanation and prediction. The further away the stars are as sources of light energy, the more energy is lost by photons (tired light) in traveling to the Earth. For star distances larger than certain values, the energy lost from visible light is large enough so that the red shift and loss of light photon energy brings it below the range of visibility. In fact the energy can drop into the microwave, and infrared range, and radio frequency ranges, which can be seen only by special detectors. This can result in the observed cosmic microwave background, CMB. For the vast number of stars at large enough distances, there will not be enough energy for light to reach us in the visible range, and the sky will be dark.

Chapter 27

Cosmic Microwave Background (CMB) Explained

The cosmic microwave background and its measured temperature of 2.7 K degrees are frequently advanced as a proof of the big bang.

We will show how the microwave energy can result, and also explain how this radiation reaches and equilibrates at this low temperature, without supporting the belief in the big bang.

When photons from remote stars and galaxies travel to the observer, the photons lose energy (due to gravitational effects) and the wavelength is increased. The photon energy drops from the visible range of light, down to the infrared, then to the microwave frequency of the X-band, about 3 centimeters, and then to the radio frequency range (meters).

This reduction of photon energy is associated with photon travel involving large distances, and provides light energy from all the stars within the event horizon associated with the observer.

A question is why the microwave frequency in the X-band, about 3 centimeters, above the energy of radio frequencies, can be observed and to be at the low temperature of 2.7 K degrees.

The temperature of space has been independently been calculated by many others to be in the low temperature range of about 5 K degrees. It involved calculations using estimates of the available energy from visible stars. (Assis,

1995) (See the History of 2.7 K temperature in the section for References and Reading Material in this document)

As a result of the gravitational interaction of photons with gas and dust in the interstellar space (tired light), photons will come into energy equilibrium with the low temperature of gas and dust in the interstellar space. This does not involve collisions or absorption and reemission, which would blur the normally sharp images of remote stars and galaxies.

To explain the process of energy transfer from photons when passing by gas and dust in interstellar space, we must consider that the photons also have mass which permits them to interact gravitationally with the low mass of gas and dust. This gravitational energy transfer is more effective when the gravitationally interacting masses have low masses closer to the photon masses.

The energy transfer from the photons is similar to the transfer of energy from our moon to the earth through gravitational drag on tides. The moon loses energy, its velocity decreases, and its orbit diameter increases.

Some of the photons will not pass close enough to the gas and dust to interact effectively and can continue to lower energies characteristic of the longer wavelength radio frequencies.

For cases of effective encounter of photons and gas and dust, the photon energy does not drop below that of the gas or dust because in this case the encounter pulls the gas or dust towards itself as it approaches (giving kinetic energy to the encountered material) and because of the effective gravitational coupling, the encountered material will share and return some of its energy to the photon to reach thermal equilibrium at the temperature of the gas and dust. The effectiveness of the encounter depends upon the distances between the photon and gas and dust, as well as the relative masses. Energy transfer and equilibrium is more effective for close distances and when the masses are close in value.

This is an alternate explanation for the Cosmic Microwave Background, CMB, and may be superior to the supposed cooling of the high initial temperature of the speculative big bang.

The classic CMB map charts the temperature fluctuations for various locations in the sky. These fluctuations are orders of magnitude smaller than the local average, and are believed by others that these fluctuations support the theory of the big bang. I have not seen any explanations for the supposed support. As we have explained elsewhere, the red shift does not support the receding galaxies or the big bang because it is only largely due to gravitational effects. We need another means to explain the tiny temperature fluctuations, but they could be due to relatively tiny differences in the travel distances.

I suggest that the temperature fluctuations are probably due to the tiny differences in the spacing of the galaxies and stars, and the gaps in the structures in the webs of galaxies, resulting in the tiny differences in the photon travel (compared to the total travel) and the distribution of relatively tiny interstellar voids in the photon path. Because the photon energy loss during travel depends upon distances traveled and the gravitational interactions with gas and dust in the interstellar voids, there will be fluctuation in the photon temperature (energy) distribution that can explain the observed temperature fluctuations.

Also the CMB has areas of observed polarization, and can be explained by the fact that galaxies including spiral galaxies consist of rotating plasmas containing free electrons and ions with a net charge (probably positive ions), producing magnetic fields that will cause polarization of the electromagnetic fields when they are scattered by the electrons. The orientation of the magnetic field and the resulting electromagnetic polarization will depend on the orientation of the axis of rotation of the galaxy.

The belief in the CMB discovered in 1964 is interpreted as proof of the big bang, and persists to date as evidenced in a recent article in the May 1, 2009 pp. 584-586 issue of the respected Science magazine. Note that a Nobel Prize was awarded to the CMB discovers.

Subsequent extensive and expensive spacecraft measurements including the 1992 NASA Cosmic Background Explorer (COBE) measured the microwave spectrum and detected tiny radiation temperature variations of about one part in 100,000s across the sky view. Indeed, this observational work won a Nobel Prize as recently as in 2006.

The Wilkinson Microwave Anisotropy Probe (WMAP) in 2003 made detailed charts the temperature variations.

Continuing work to study the CMB includes Europe's Herschel Space observatory to fill in the electromagnetic region between the far infrared and submillimeter wavelengths between 200 and 850 micrometers.

The European Space Agency (ESA) will launch the Planck satellite to map in greater detail the patterns in the polarization of the microwaves.

The value of these important observations and measurements is reduced by the resulting speculations and theories that use them to support the speculation about the big bang, the early inflation, and the supposed and incorrect makeup of the universe – as described by others as apparently consisting of 4% ordinary matter 23% Dark Matter, and 73% Dark Energy related to the supposedly expanding universe.

We have attempted in other chapters to show that Newton's law of gravity in not a universal law that also can be used wrongly for cosmic distances to support the need for Dark Matter. Also we have pointed out that Hubble's law for red shifts is only based upon distances and that expanding velocities were not measured. Also we pointed out that the linear Hubble law can not be extended indefinitely otherwise it would require decreasing photon energy to go to zero and then become negative as distances increases. This argument should eliminate the speculation and support for Dark Energy.

As an experimental physicist there is admiration and respect for data obtained by hard work, and I believe that the value should not be diminished by speculation and theories that try to use the data for support of wrong theories, leading to the standard model of the universe based upon misunderstandings.

I have personally and gratefully benefited from over five years of continued sponsorship from NASA Headquarters that permitted me to work on plasma and Magnetohydrodynamic propulsion theories and systems for space vehicles. My PhD theses involved plasmas and about half of my work involved physical electronics, instrumentation, and plasmas so that I have familiarity with the associated physics.

Chapter 28

Black Holes And Information

We will reconsider the problem of black holes and the need for the conservation of information when black holes evaporate and disappear. The well-known and brilliant Stephen Hawking has described the black holes (Hawking, 1988).

There are some questions and uncertainty about the loss of information connected with material drawn into black holes. Even Stephen Hawking has recently changed his ideas about black holes.

Proposed here is a simple theory for the evaporation that gradually reduces, evaporates, or eliminates black holes and that preserves and releases the information collected and contained. It is consistent with General Relativity in terms of gravitational attraction, and with Quantum physics in the form of nuclear processes and thermodynamics for photon production and emission. We will try to avoid speculation and will draw on accepted concepts in physics.

In my analysis of the physical processes, there is encountered the suggestion that there is a need for preservation of information drawn into the black hole.

I suggest that the mass and information in the black hole can evaporate in the form of emitted phonons from the high temperature stars trapped and radiating in the back hole.

We are all familiar with information in the form of photons. They are involved in transmitting information from books, scenery, and television and computer screens.

The mass trapped in the black hole leaves the hot stars through the nuclear conversion of mass into high temperatures and the resulting photon energy distribution. This photon energy is in thermal equilibrium at a high temperature with a black body Maxwell-Boltzmann energy distribution. Thus there are always very high-energy photons in the high-energy tail of the distribution, and after they are emitted from the black hole they are replenished from the body of the thermal energy distribution. The high-energy photons with sufficient energy in the tail of the thermal distribution can leave by escaping over the gravitational energy barrier of the black hole and lose energy in the process.

Stars and other material (gas, dust) drawn into a black hole will reach a high sun temperature and will slowly leave in the form of photons, The massive gravitational well associated with the black hole contains information which is said to need to be preserved.

The information content in black holes leaves in the form of photons. The information about the universe reaches us in the form of photons and electromagnetic energy, thus information in some form is available and preserved in the photons emitted from the black hole. For example, information frequently exists in the form of images provided by light (electromagnetic form).

Some of the photons emitted from the stars just at the edge of the black hole, entering the gravitational well, and slowly spiraling down, will have enough energy to overcome the gravitational energy loss needed to escape and will emerge with reduced photon energy and at longer wavelengths. This adds to the evaporation from the black hole.

I suggest that the radius of the black hole will increase when observed by photons with longer wavelengths, such as radio frequency rather then in the visible range.

Other stars further down in the gravitational funnel and spiraling down over millions of years will still produce photons with a black body spectrum. In the high-energy tail of the spectrum, there are always photons energetic enough to overcome the deeper portions of the gravitational well and can evaporate, thus reducing the mass and information content of the black hole. The thermal equilibrium in the stars will act to replace the high-energy photons that will continue to escape. The nuclear process in the stars will continue to convert matter into photon energy, and thus the escaping photons will be equivalent to the escape of matter and information from the black hole.

The photons from deeper in the gravitational well will lose enough energy so that the exiting photons will not be in the visible range, thus supporting the term "black hole", however the black hole will be visible (probably with larger diameter) when viewed at longer wave lengths such as infrared, microwave, and radio frequencies.

Matter (including dust and gas) not already radiating photons will be drawn down to the center of the black hole where the gravitational compression can increase the temperature and nuclear processes - thus producing additional black body radiation, and thus continuing the escape of energetic photons and of mass converted into photons.

The net rate of evaporation of the black hole is determined by the rate of replenishment by nearby matter drawn into the black hole. When the neighborhood is depleted enough so that the rate of replenishment does not replace the rate of photon-mass evaporation, the mass and gravitational trap of the black hole decreases, thus reducing the attractive force to drawn in more matter, and the photon-mass-information rate of evaporation increases - similar to the process of positive feedback.

At some point in the future the residue will probably consist of a few stars including dense neutron stars, and the massive black hole and strong gravity has departed, The associated information has also left.

Other alternate suggestions including quantum explanations apparently are not needed. Also, Multiple universes as recently suggested by Hawking apparently are not needed.

Theories and speculations involving singularities are to be considered suspect and apparently are signaling the presence of errors.

Chapter 29

Space Time Concept

Science resides in several forms.

One form is in the real world, and is a physical form that uses observations to describe, understand, and predict physical manifestations of the universe.

The other form is in the theoretical or mathematical form where equations and theory are used to explain, describe, understand, and predict results in the physical form

The validity of science in the theoretical form depends upon its ability to describe the current observations and future observations in the physical form, and to do this within the accuracy of the observations. The real world of observations must take priority over the results of the theoretical form if there is no agreement.

They differ in the sense that in the physical form the observations need to be reliable so that they can be repeated with the same or approximately same results. Preferably they can be repeated and independently verified by others, perhaps at different locations and with different equipment.

In the theoretical or mathematical form, there is much more freedom to propose various theories, but they must pass the test of corresponding to valid observations in the physical world.

There are many current theories including parallel universes, wormholes, and speed faster than light.

One respected theory by Albert Einstein is important and involves space-time with four dimensions, and is related to the three spatial dimensions plus the time dimension. One aspect to this theory is the relationship between the distortion of space-time by mass and the resulting gravitational attraction.

Actually the concept of space-time is consistent with many observations in the physical world, but there may be other theories that also achieve the same agreement.

It is very easy to add theories, even if they are not expected to correspond to reality. A simple computer program can be written to automatically supply many additional apparently plausible theories by modifying current theories by scanning each word and substitute for some suitable words with randomly chosen substitutes from the same type such as verbs, adjectives, or nouns. A simple way is to negate words such as "is" to the word "is not", or, the word "big" to the word "small", or the word "fast" to the word "slow", or the word "white" for "black" as examples. We can have all the theories we want.

Remember, observations in the physical world must take priority over just theories. If the theory does not agree with valid observations, the theory is not sufficient and must be abandoned or modified.

Use of space-time as an explanation for the observed gravitational effects is just a theory to help understand observations. They do not control observations.

According to Einstein's general relativity, gravitational fields are described in terms of the geometric curvatures of space-time.

However, the theory of the curvature of space-time as being associated with gravity and influencing the motion of bodies, does not result in explanations for the mystery of Dark Matter.

We suggest that the limitations of the theoretical model of space-time should be studied.

Chapter 30

Virial Theory in Astrophysics

The Virial theory can be used to describe the collective action of groups of particles that interact gravitationally. It involves a theoretical and mathematical relationship between the time averages of the total potential energy and of the total kinetic energy.

However, the calculation of the total potential energy involves Newton's law of gravity and the gravitational constant. For particles with cosmic separations, the use of Newton's law for much smaller distances (as in our solar system) there will be problems in the use of the Virial theory.

The rest of this chapter is very mathematical and need not be read once you understand that it assumes Newton's gravitational constant and in order to be accurate for galaxies must use the TAG version of Newton's constant that is valid at cosmic distances.

The time average of the total potential energy of the group is $<T>$, and the time average of the total kinetic energy is $<V>$.

The relationship of the two averages is $<T> = <V>/2$ and the involved mathematical details of the derivation are available in the book, "Gravitational physics of stellar and galactic matter", by William CZ. Saslaw, Cambridge University Press, Cambridge, 1985

When Fritz Zwicky, in the early 1900s, used the Virial theory in conjunction with his analysis of the observed behavior of groups of galaxies he found that the Virial theory could not fit the observations unless there were massive amounts of missing matter, much greater than the visible matter, to provide the necessary gravity to fit the theory. This apparently was an early introduction of the concept of Dark Matter. However, at that time it was believed by many (or all) that Newton's law of gravity was a universal law

My explanation of the failure of the Virial theory and equations to agree with the observations is that the time average of the potential energy $<T>$ incorrectly used Newton's gravitational constant G in calculating the potential energy. We have shown that for cosmic distances, the gravitational constant should be extended (not replaced) by the theory of additional gravity TAG involving the addition of a linear component in order to be universally valid also at cosmic distances.

My analysis of the observed motion of stars in spiral galaxies by Vera Rubin, (Rubin, 1970, 1985) plus an elementary equation balancing the inward gravitational force with the outward centrifugal force of rotation, showed that for the region where the observed rotation velocities were constant, it required that the product of the mass, M, and Newton's gravitational constant, G, should really be a linear function of distance, r.

Rather than the usual assumption that Newton's law is also valid at cosmic distances, which requires that the mass M to have an invisible component that increase with the distance, we introduce our solution that the gravitational constant supplies the linear requirement. This addition to Newton's gravity constant eliminates the need for the need for the extensive and failing search for invisible Dark

Matter. Remember that gravity itself is already invisible and the addition of an invisible linear term should be logical.

Another important contribution by Zwicky was his suggestion of the concept of "tired light" where the energy of traveling photons was reduced by gravitational interactions. This is an explanation of the linear dependence of the red shift on the distance traveled.

Others have objected on the basis of loss of photon energy through collisions or absorption followed by emission would result in change of direction and fuzzy images. However only sharp images produced by red shifted photons are considered, and not any fuzzy images.

In my opinion Fritz Zwicky was very knowledgeable and contributed much to the information about the universe. Apparently others did not respect his contributions and did not take him seriously. Such rejection of important information could have made him frustrated and angry.

Chapter 31

An early essay on the universe
By Sol Aisenberg

This information is provided in chapters that can be read without significantly depending upon information in other parts of the document, and thus many chapters will contain redundant information.

Gravity, which is not really understood, including its cause or origin, influences all of us every moment. It is also a key factor in the formation of stars because the stars are ignited by the energy of gravitational collapse of mass, leading to the nuclear processes for continued conversion of mass to energy. Gravity also influences the energy and red shift of the photons that provide us a view of the universe. Unless we understand the roles and contributions of gravity in the universe, it will be difficult to obtain a model of the universe that does not have the many current mysteries.

This analysis describes my Theory of Additional Gravity, (TAG), for the Universe. It agrees, without change, with the gravitational theory of Newton and of Einstein for smaller separations such as in our solar system, and it is also valid and significant at galactic separations.

Presentation of my Theory of Additional Gravity (TAG) can simplify the model of the universe. In order to prevent this theory from being classified as speculation, items of supporting information based upon observations published by others will be included. Identification of other beliefs will be identified as consisting of speculation where this is the case. A number of my comments about other consequences will be included.

My model is actually different from the interesting MOND theory of M. Milgrom involving modified gravity. (Milgrom, 1983, 1998, 2002) It does not involve the MOND model, which uses acceleration plus a non linear dependence and interpolation between limits.

My suggestion is that many of the beliefs about the universe are wrong and needlessly complicated and use the invocation of mysterious and massive fudge factors such as Dark Matter and Dark Energy. I need only one clarification of the effect of gravity, and one clarification of the true meaning of the red shift to start to simplify the model of the universe. They are consistent with observations. Of course, I take responsibility for any of my errors.

The proposed model of the universe does not need to involve quantum theory or Einstein's Relativity, nor does it question them. They are left to the many scientists who use them to explain many of the processes in the universe. Before attempts are made to unify the theory of gravity with quantum theory, one should be sure that the correct model of gravity is known and used.

However Einstein's theory of General Relativity involving the effect of gravity on photons of light is involved and is important in explaining the additional gravitational contributions to the red shift.

In our work, observations and experiments must take priority over theory, particularly when they are reproducible and when there are conflicts with theory.

For over 70 years (starting in about the 1930s) the scientific community has been concerned about additional problems and mysteries in the understanding of the universe. One problem is finding Dark Matter, many times larger than the visible matter represented by light from stars. Other problems are the apparent expansion of the universe and the apparent acceleration of the expansion, and the postulated Dark Energy and negative gravity.

According to my analysis, the mysteries and problems are caused by two fundamental ASSUMPTIONS made by others, which are commonly used without proof.

One ASSUMPTION of others is that Newton Law of Gravity is a Universal Law, without modification, or proof of universality, and while valid in our solar system (with a slight puzzle – the Pioneer Anomaly) is also assumed to be valid at very large distances outside our solar system. Actually we can show, with a simple equation, and based upon observations published by others, that the gravitational constant includes a term that increases linearly with distance, and that this effect is readily apparent at cosmic distances.

The second ASSUMPTION of others, again made without proof, is that the red shift and the Hubble constant can be used to measure the velocity of remote stars and also the distance for very far distances. Actually the red shift is based upon observations involving distance and not velocity. It cannot be used to measure velocity, but only of distance. The receding velocities of the apparently receding galaxies were not directly measured.

In fact, we will show that in certain cases involving extreme distances the red shift cannot even be used as a measure of distance.

This incorrect assumption about the red shift as a measure of receding velocity is serious because it has produced a false belief that the universe is expanding, that the expansion is accelerating, the need for inflation, and that there is a need for Dark Energy.

Associated with the assumption of the rapidly expanding universe is the concept of the big bang, supposedly and partly supported by the Cosmic Microwave Background (CMB) radiation and the observed temperature of 2.7 K degrees for the radiation, and also the uniformity of the CMB.

Also part of the standard theory is the apparent need for the concept of the inflation phase of the beginning of the universe to explain the observed uniformity of the universe beyond the limit of the range limited by the velocity of light and the assumed age of the universe, (the event horizon).

The event horizon based upon the estimated life of the universe and the velocity of interaction limited to the velocity of light, C, requires inflation as part of the big bang theory to explain the observed uniform nature of the universe.

Our new model of the universe also explains the cosmic microwave background and the uniform radiation temperature without requiring a big bang or inflation.

To differentiate our new simplified model from pure speculation, we will provide a number of items that provide supporting arguments for our new theory. They are based upon the vast body of astronomical observations reported by others - because I do not need or have access to the excellent observational equipment used by them.

As part of my contribution of the my new theory, I will identify and suggest some tasks that can be done by others who have access to primary astronomical data, or who can carry out calculations to evaluate the suitability of my theory.

Initially, starting in 1998, I became aware of the observation of the constant velocity rotation curves of spiral galaxies as reported by Vera Rubin and others. This lead to my determination that Newton's gravitational theory and the gravitational constant, G, has an additional gravitational term that increases linearly as a function of separation, r. It provides a simple extension of the gravitational force of Newton and Einstein that is only significant for large separations and that, without modification for small distances compared to cosmic distances, reduces to Newton's law in our solar system. Unexpectedly it leads to the understanding and correction of many other mysteries in the standard model of the universe.

In the beginning of my obsessive, part time research on this subject (in about 1998), my new theory was that Newton's gravitational constant, G, can be generalized and expanded into a simple power series in terms of distance, r, and could explain the observed flat rotation velocity of stars in spiral galaxies as reported by the capable Vera Rubin. In my initial simple approach, it used the first linear term in the power series expansion of G with respect to distance, r, and the gravitational constant Ga with the additional term was now Ga = Gn + A*r where Gn is Newton's gravitational constant, r is the distance, and where A can be proven to be non zero when evaluated using published observations of spiral galaxies. When asked where the term A*r comes from, an answer can be that it comes from the same place as Newton's gravitational constant.

Thus according to my new theory, the inverse square attractive force between masses is augmented at very large separations by another component that decreases much slower than the usual inverse square dependence on distance. This involves an additional term in describing the gravitational constant.

Further recent study showed that there is no need for a power series expansion. The new representation of the long range gravitational constant Ga can result from the elementary physics equation describing the equilibrium between the gravitational force, $G*M/r*r$ towards the central mass M, and the centrifugal force of rotation, $v*v/r$ for stars with velocity, v.

The resulting equation is now $M*G = v*v*r$, and for the observed case of constant rotational velocity, v, it reduces to $M*G$ as a linear function of r.

The usual interpretation and assumption by most others is that G is assumed to be independent of distance, and therefore the mass M must provide the necessary linear dependence. The unproven halo of Dark Matter around the spiral galaxy was the result, where the invisible, dark, mass was computed to be much larger than the visible mass based upon the observations. Note that the supposed Dark Matter also had the properties of not emitting or reflecting light, and of not eclipsing other light, but was only evident through its gravitational force.

In addition to explaining observations at cosmic distances, this new linear contribution to the gravitational constant, without change, reduces to the Newton and Einstein's description of gravity in our solar system where separation distances, r, are much smaller values than galactic separations. The gravitational theories of Newton and Einstein are not replaced but just extended.

The additional gravity is too small to be measured for the relatively smaller distances in our solar system - with the exception of the very precise measurements possible using the NASA space probes, Pioneer 10 and 11. Their long-term precise measurements using modern techniques detected a Pioneer Anomaly corresponding to very tiny additional forces towards the central sun. This anomaly alone in our solar system shows that the theory of classical gravity should be reexamined and extended, as we have done.

Others have suggested a modification of Newtonian gravity. One significant suggestion was MOND (Modified Newtonian Dynamics) hypothesized in 1983 by Moti Milgrom. Briefly, in the MOND version the modification of the effect of gravity occurs at very small accelerations, involves nonlinear acceleration terms, and uses interpolation functions.

It is different from our simple Theory of Additional Gravity (TAG), which involves generalization of the gravitational constant that derives an additional linear term from an elementary equation that depends upon separation and does not depend on acceleration.

As another aspect of our analysis of the standard model of the universe, we determined that the meaning of the red shift shows that it can only measure distance and not velocity for remote stars and galaxies. We identified three ways that the red shift has contributions provided by gravitational effects alone, in addition to any Doppler effects. In fact in certain cases where black holes are present, the red shift cannot be accurate in determining distances because of the effect of massive gravity in reducing photon energy and increasing the red shift.

Also, for vast distances, the linearity of the Hubble relationship must saturate and become non linear - otherwise it would predict for very large distances the unlikely need for the photon energy to become zero and even negative.

Implications of our studies can influence and explain the supposed expansion of the universe, the acceleration of the expansion of the universe, negative gravity, Dark Energy, the big bang, and the surprising deduction and computation of transverse velocity of very remote galaxies being greater than the velocity of light.

An important part of the value and validity of a new theory is the ability to agree with and explain existing observations, but even more important to make many predictions that can be confirmed by future observations.

This new analysis can encourage capable scientists including physicists to redirect their valuable time away from the fruitless research and search for Dark Matter, and into more important problems.

Chapter 32

Another early essay on the universe
By Sol Aisenberg

Much of the information provided in the various chapters will repeat material in other chapters because this is a collection of chapters that were prepared over periods of time, and expanding on new insights and different ways of explaining my analysis.

It is probably useful in repeating the important material in different ways to insure that the reader can absorb the material and new concepts.

Also, it may not be necessary to read and study all the chapters, but the reader can focus on the chapters of particular concern before perhaps studying all the material.

This analysis describes my theory of additional gravity for the Universe that agrees with the gravitational theory of Newton and of Einstein, for smaller separations such as in our solar system - and without changes, is also valid and significant at galactic separations.

Only a few basic equations are needed and used, so that the average reader without physics training also can understand the material and explanations. I have learned from experience that when answers to problems are finally obtained, they frequently are found to be simple and apparently surprisingly obvious afterward.

Also when a theory involves singularities or infinities when taken to the limits, this tells me that the theory needs questioning and revision, and to be used with care. This is the case for the current standard model of the universe, which involves an initial theoretical singularity.

Presentation of my Theory of Additional Gravity (TAG) can simplify the model of the universe. In order to prevent this theory from being classified as speculation, items of supporting information based upon published observations by others will be included. A number of related comments will be included.

In order to help establish my credentials as an educated, productive, and experienced scientist, support is provided in more detail as Appendices at the end of this document. Most of my work has been in industry because this gave me the opportunity to contribute in many scientific fields, depending upon my ability to obtain funding for myself and for my associates, although part-time I participated in academic activities when I could spare the time.

About half of my work, publications, reports, and presentations were in the area of plasmas, materials, energy conversion, optics, and instrumentation, while the remainder was in the area of medical technology. The distribution largely depended upon the availability to obtain funding.

My model of the universe is really different from the interesting MOND theory of M. Milgrom involving modified gravity. It does not involve the MOND model, which uses acceleration plus a non linear dependence and interpolation between limits.

My suggestion is that many of the beliefs about the universe are wrong and needlessly complicat00ed and use the invocation of mysterious and massive fudge factors such as Dark Matter and Dark Energy. I need only one clarification of the effect of gravity, and one clarification of the red shift to simplify the model of the universe. They are consistent with observations. Of course, I take the usual responsibility for any of my errors.

The proposed model of the universe does not need to involve quantum theory or Einstein's Relativity. They are left to the many scientists who use them to explain many of the mysteries of the universe. However Einstein's theory of General Relativity involving the effect of gravity on light photons is involved and is important in explaining the additional gravitational contributions to the red shift.

The current attempts to reconcile the theory of gravity with quantum theory will probably fail until the correct theory of gravity is established and used.

In our work, valid observations and experiments take priority over theory, particularly when conflicts exist.

For over 70 years (starting in about 1930) the scientific community has been concerned about problems and mysteries in the understanding of the universe. One problem is finding Dark Matter, many times larger than the visible matter represented by observed light from stars. Other problems are the apparent expansion of the universe and the apparent acceleration of the expansion, and the postulated Dark Energy and negative gravity.

According to my analysis, the mysteries and problems are caused by two fundamental ASSUMPTIONS of others in the field, which are commonly used by others without proof.

One ASSUMPTION made and used by others is that the Newton Law of Gravity is a Universal Law, without modification, or without proof, is that it is also valid at very large cosmic distances outside our solar system. Actually we can show, based upon observations of others, and a very elementary equation, that the gravitational constant includes a term that increases linearly with distance, and that this effect is apparent at cosmic distances.

The second ASSUMPTION of others, again used without proof by others, is that the red shift and the Hubble constant can be used to measure the velocity of remote stars and the distances for very remote galaxies and stars. Actually the red shift is based upon observations involving distance and not velocity. It cannot be used to measure velocity, but only for distance. In fact, we will show that in certain cases of large distances it cannot even be used to measure distance.

This incorrect assumption about the red shift as a measure of receding velocity is serious because it has produced a belief that the universe is expanding, and even fooled Einstein. It apparently also showed that the expansion is accelerating, and that there is Dark Energy.

Associated with the assumption of the rapidly expanding universe is the concept of the big bang, supposedly and partly supported by the Cosmic Microwave Background (CMB) and the observed temperature and uniformity of the CMB.

Also part of the standard theory is the concept of the inflation phase of the beginning of the universe. The event horizon based upon the estimated life of the universe and the velocity of interaction limited to the velocity of light, C. It requires inflation as part of the big bang to explain the observed uniform cosmic microwave background and the uniform temperature. Our simplified model explains the cosmic microwave background and the uniform temperature without requiring a big bang or inflation.

To differentiate our new simplified model from pure speculation, we will provide a number of items that provide supporting arguments for our new theory. They are based upon the vast body of astronomical observations already reported by others - because I do not need or have access to observational equipment.

As part of my contribution of the my new theory of additional gravity TAG, and the resulting consequences, I will identify and suggest some tasks that can be done by others who have access to primary astronomical data, or who can carry out calculations to evaluate the suitability of my theory.

Initially, starting in 1998, I became aware of the observation of the constant velocity rotation curves of spiral galaxies as reported by Vera Rubin and others. This lead to my determination that Newton's gravitational theory and the gravitational constant, G, has an additional gravitational term that increases with separation r. It provides a simple extension of the gravitational force of Newton and Einstein that is only significant for large separations and, without modification, reduces to Newton's law in our solar system. Unexpectedly it leads to the understanding and correction of many other mysteries in the standard model of the universe.

In the beginning in 1998, my new theory was that Newton's gravitational constant, G, can be generalized and expanded into a simple power series in terms of distance, r, and could explain the observed flat rotation velocity of spiral galaxies. In the simple form, the gravitational constant Ga with the additional term is now $Ga = Gn + A*r$ where Gn is Newton's gravitational constant, r is the distance, and where A can be proven to be non zero when evaluated using published observations of spiral galaxies. When asked where the term A*r comes from, an answer can be that it comes from the same place as Newton's gravitational constant.

Thus according to my new theory, the inverse square attractive force between masses is augmented at very large separations by another force that decreases much slower as a function of separation and involves an additional term in describing the gravitational constant.

Further study showed that the new representation, (now not involving a power series expansion or approximation), of the long range gravitational constant Ga can result from the elementary physics equation describing the equilibrium between the gravitational force, G*M/r*r towards the central mass M, and the outward centrifugal force of rotation, v*v/r. The resulting equation is M*G = v*v*r and for the observed case of constant velocity, reduces to M*G as a linear function of r.

The usual interpretation of the situation is that G is assumed to be independent of distance, and therefore the mass M must provide the necessary linear dependence. The unproven halo of Dark Matter around the spiral galaxy was the result. Note that the supposed Dark Matter had the properties of not emitting or reflecting light, and of not eclipsing other light.

This linear contribution to the gravitational constant reduces, without modification, to the Newton and Einstein's description of gravity in our solar system where separations approach smaller values than galactic separations.

The additional gravity is too small to be measured in our solar system with the exception of the very precise measurements possible using the NASA space probes,

Pioneer 10 and 11. Their long-term measurements detected a very tiny additional force towards the central sun. This alone shows that the theory of classical gravity should be reexamined.

Others have suggested a modification of Newtonian gravity. One significant suggestion was MOND (Modified Newtonian Dynamics) hypothesized in 1983 by Moti Milgrom. Briefly, in the MOND version the modification of the effect of gravity occurs at very small accelerations, involves nonlinear acceleration terms, and uses interpolation functions. It is different from our simple Theory of Additional Gravity (TAG), which involves generalization of the gravitational constant that derives a linear term from an elementary equation that depends upon separation and does not involve acceleration.

Analysis of the meaning of the red shift shows that it can only measure distance and not velocity for remote stars and galaxies. Gravitational effects influence the red shift. Implications of our studies can influence and explain the supposed expansion of the universe, the acceleration of the expansion of the universe, negative gravity, Dark Energy, the big bang, and the deduced transverse velocity of very remote galaxies at velocities greater than the velocity of light.

An important part of the value and validity of a new theory is the ability to agree with and explain existing observations, but even more important to make many predictions that can be confirmed by future observations.

Chapter 33

Some Tasks For The Future

Curiosity is frequently a driving force for scientists and others, and is largely responsible for the advances leading to the current civilization.

Over centuries people such as Copernicus and Galileo took personal risk to share the knowledge that their curiosity obtained. Others searched for knowledge because of their desire for fame and money. Currently scientists and engineers are dependent upon funding for their continued work and income. This limits the direction of the current activities because it must correspond to the wishes of the sources of funds. In the past individuals who were wealthy were able to build telescopes and to learn about as much of the universe that could be seen. Einstein was free to pick the direction of his massive contributions because he had income from his patent office job.

Under the present conditions, scientists must spend much of their time in writing proposals for funding for their work and for support of their students who are working toward their degrees. If a student selects a topic for his degree that is out of favor or that is seriously in conflict with the common scientific beliefs, he or she is in danger of not receiving the degree, except for the rare cases (such as for Shannon and information theory) where the results are so powerful that the material is quickly accepted.

I suggest that much of the scientific progress in the future will come from those who are funded by individuals, who are philanthropists, or from retired scientists, or semi retired scientists, who have earned their scientific freedom.

In my past work, supported by contracts and funding from many federal agencies, there was a need to work on directions specified by the request for proposals (RFP), and I appreciated the support. This support permitted me to also work separately on problems that I was able to select. As a result I became more productive in many disciplines. My natural desire and interest in different problems provided the ability to solve problems, with the advantage of fertilization of knowledge from other fields.

It would be nice and beneficial if the government would provide greater funds with fewer strings for scientists and people who demonstrated scientific potential so that they are free to chose the problems that they feel are important. Progress is hard to make if you are forced to follow in the tracks of others and not free to branch off in promising directions.

One solution is to start your own consulting, inventing, and licensing company (as I have done) so that when it is successful you are free to select your own scientific problems and other national problems to work on.

We have the resources consisting of many scientists and engineers, and they should be funded, redirected, and expanded to solve the many national and international problems that affect humanity.

Note that some of the problems are not scientific but it is important that successful scientists, inventors, and business people use some of their vast funds and political power to counter and limit the actions of the lobbyists funded by companies who may be controlling Congress for the benefit of companies rather than for the benefit of citizens, taxpayers, and voters

Scientists and engineers can solve some of these problems, and some problems need political power and funds that can be supplied by federal funding, foundations, successful business people, and successful scientists and technologists

The following are some suggestions for problems to be considered:

Energy

* Finding new practical sources of energy

* Finding new replacements for petroleum

* Devising better storage batteries

* Solving the problem of convenient room temperature superconductor or higher conductivity wires for the electric power grid

Health

* Providing means to control or limit viral/bacteria epidemics

* Reducing the cost of health care

* Creating new and better means to improve the health of people

* Health care and insurance for everyone and that is more effective and with control of costs

* Needed are more effective ways of controlling or curing cancer

* Better and less expensive ways of controlling infections, such as bacteria and viruses

* Devise improved stents to improve blood circulation

* Devise practical way of assisting hearts while solving the limitations of tissue and blood compatibility

Water

* Solving the need for new sources of water

* We will need to provide water for the parched regions, possibly by desalinization of ocean water, using unlimited energy

Universe

* Arranging to detect and prevent asteroids from impacting the earth with enough force to destroy most life, as has happened in the past.

Survival

* Recognizing and determining if significant global warming really exists

* Providing means to survive global warming if it is true

* Realizing and showing that the sun itself has actually determined the temperature of the earth in the past thousands of years, and before the effects of man

* Showing that CO_2 may be accelerating the warming of the earth but realizing that reducing this increase in CO_2 will only delay the warming for only the present generation.

* Sequestering the excess CO_2 in plants as a more efficient way of recycling it together with the sunlight stored in plants, for a future energy resource as was done by nature in past eons

* We will need essentially free energy to power the means to permit us to survive the higher and lower temperature periods

* We will need crops that are more productive and efficient, and free of patent control by agricultural seed companies.

General

* Better ways of seeing that members of congress represent the true wishes of the country rather than being controlled by lobbyists

* Devise ways of controlling hackers and crimes on the Internet

* Improving the means for transportation

* Reducing crime

* Improve the Internet defense capabilities of companies

* Providing survival in the event of significant global warming or cooling

And many other national problems that the reader can identify

Chapter 34

Conclusions

The currently accepted model of the universe is accepted by the majority of experts consisting of physicists, scientists, cosmologists, and politicians, in spite of the many mysteries such as Dark Matter and Dark Energy.

In past centuries it was dangerous to oppose the common beliefs about the solar system, and it took brave people like Copernicus, and Galileo to provide their new insight.

In later centuries, Newton and Einstein were able to provide their contributions to knowledge without risk of final rejection. They did this in spite of the limited astronomical observations available to them, and used their intelligence to compensate for the lack.

In the recent generations, starting in about 1900, with the explosion of technology and observing capability, there were many more, higher quality observations of the universe available. About that time, there started to be a number of wrong conclusions and mysteries connected with these new observations, many of which were for observations at cosmic distances. The science of cosmology outside our solar system is unique in the sense that it mostly depends upon information that arrives at the observers only in the form of light and other electromagnetic forms. It is not subject to confirmation by experiment, but only by repeated and possibly more accurate observations.

The science of cosmology is now in a state where the standard model of the universe needs careful review and corrections. Otherwise, the scientists and engineers working in the field may find that they have been wasting their precious time that cannot be recovered. The wasted sponsorship money can be replaced in the future, provided that it is not needed for other immediate research.

As a researcher, I strongly support research, even if some of it turns out to be of less value, because it can be of benefit for educating and training future researchers. Remember, research is important and necessary for the existence, future, and progress of civilization.

Enough information has been presented in the various chapters of this document, that future generations, (and possibly some in the present generation), can have a start in arriving at better, more accurate, and believable version of the misunderstood universe.

Chapter 35

Summary

We have considered some of the more important, and incorrect aspects of the misunderstood Standard Model of the Universe and suggest some corrections using our analysis.

Included are discussions of: Dark Matter, the gravitational constant and additional gravity, the Red shift and tired light, the Hubble constant, the Age of universe, Dark Energy, red shift effect on the apparent motion and energy of Galaxies, the big bang, and inflation.

In our opinion, valid observations must take priority over theory when the theory is not in agreement with observations.

Speculation can be the beginning of understanding but should not exist in valid solutions or explanations.

An indicator of speculations can include any of the following words, for example: apparently, suppose, suggest, perhaps, maybe, or similar phrases.

My personal belief is that science is based upon observations. Whenever the theory does not agree with observations (validated by repetition, and preferably by others) then the observations take priority. Then the theory in order to be considered must be modified or extended in order to be valid. Theories however, have value because they can guide the attempted validation by new observations.

Dark Matter

Newton's law of gravity is based upon extensive data for the motion of bodies in our solar system. For observations at cosmic distances outside our solar system, the law fails. The concept of Dark Matter is needed to make the cosmic observations fit Newton's law. This is the case for the Rubin's observed flat rotation velocity curves of spiral galaxies and the observed motion of groups of galaxies as reported by Zwicky. When we extend Newton's gravitational constant by a term linear in distance, r, then Dark Matter is not needed to explain these observations. Our solution is different from the MOND theory of Milgrom that involve acceleration.

Remember that gravity itself is invisible and our analysis and simple invisible extension can explain observations without needing Dark Matter.

One basic assumption in the commonly accepted model of the universe is the assumption that Newton's laws of gravity and the gravitational constant are also valid at cosmic distances.

Newton's law and gravitational constant were shown to be valid but only based upon astronomical observations and subsequent laboratory experiments in our solar system.

There is no observational proof that Newton's laws are also valid for the universe or for distances larger than for our solar system. In fact we find that a simple extension of Newton's original law of gravity for large distances will fit the observations at cosmic distances and also in our solar system without change. (This is different from the interesting MOND theory of M. Milgrom that involves acceleration.)

The usual assumption about gravity has led to the commonly held belief that Newton's law of gravity, combined with observations of motion of groups of galaxies, and the motion of stars in spiral galaxies need the presence of massive amounts of missing matter, now called Dark Matter to explain the many observations. This has resulted in a number of more serious errors, which are included in a number of respected references.

There are errors and assumptions in the standard model of the universe – and they should be corrected in order to really understand our universe. It is important that the scientific community take careful consideration of the correction of the assumptions in order that many capable scientists do not waste any more of their precious years of research by following ideas and trails that will be discarded by future generations.

Thus a major wrong and fundamental assumption and belief about our universe is that Newton's laws and the gravitational constant is also valid at galactic distances outside our solar system. There are no observational proofs for this assumption.

A corrected understanding of extra or additional gravity outside our solar system does not invalidate the work of Newton and of Einstein. Their work was based upon observations in our solar system and is still valid there.

Starting over seven decades ago, there now are serious errors in the commonly accepted model of the universe. Observations of the rotation of spiral galaxies together with early observations of the motion of groups of galaxies, plus Newton's laws required massive amounts of missing matter, now mistakenly called Dark Matter, to explain the many observations.

I will show how my Theory of Additional Gravity (TAG) uses a simple extension of Newton's gravitational constant Gn consisting of an additional term linear in distance r, in the form A*r, where A is the coefficient of r, can explain the observations without needing to search for Dark Matter.

Based upon the balance between the centrifugal force on a mass, m, rotating at a velocity, v, around a central mass, M, at a distance r, m*v*v*/r, and the gravitational force on the mass rotating mass, m, which is m*M*G/r*r, the resulting equation is M*G = v*v*r. For the case of the observed constant velocities, if we assume that the mass is constant, then the dependence on r is part of the gravitational constant. Thus we arrive at the following dependence for gravity with the additional linear term Ga in the form Ga = Gn * A*r.

In previous analysis we arrived at the same term Ga but by expanding Gn in a power series and used the first linear term as an approximation.

Further analysis showed that the linear term can be obtained directly from the liner dependence of M*G if we assign the linear component to Gn rather than to the mass M, thus avoiding the need for Dark Matter.

Preliminary analysis of data for rotation velocity, v, plotted against radius, r, will provide the transition radius, Rt, where the rising portion of the curve transitions to the flat portion of the velocity vs. radius plot for data from the spiral galaxies NGC2403 and NGC3198.

Using the spiral galaxy transition radius Rt of 2.7 kpc plus the known value of Newton's gravitational constant Gn, gives, for these spiral galaxies a value of A = Gn/Rt, and thus A = 1.18 x 10 exp -14 /sec*sec.

The coefficient A for r represents an additional gravitational contribution that is only significantly apparent beyond the dimensions of our solar system. (If asked where the term A*r comes from, the answer is that it came from the same place as Gn.)

$$\text{Thus Ga} = \text{Gn} + \text{A}*\text{r}.$$

Because of the elegant simplicity of this assumption compared to the long time, expensive, and unsuccessful search for massive amounts of Dark Matter, it provides a beautiful alternative.

The theory of invisible matter and Dark Matter has been used for years to explain the important measurements of Fritz Zwicky on motion of a remote collection of galaxies, and the important work of Vera Rubin and others on spiral galaxies having constant outer rotation velocity curves.

My recent hypothesis is offered to provide a simple addition to the gravitational theory of Newton and Einstein (which are still mostly valid in our solar system) to explain and predict the published observations of others, and to show that there is no need to invoke Dark Matter, and could stop the fruitless and expensive search for Dark Matter.

The need for the additional gravitational constant explains the inability of computer simulation of formation times of galaxies, voids, and other massive structures to agree with limitations of the supposed age of the universe.

The NASA space probes, Pioneer 10 and 11 show apparent gravitational anomalies in our solar system that may also be explained by the additional gravitational constant.

An objection was raised based upon the fact that Newton's gravitational law accurately explains the motion of the outer planets. However these planets do not (yet) have the instruments that would permit their motion to be measured with an accuracy of about 1 part in 10 to the 8th power, which is the case for Pioneer 10 and 11 probes.

Red Shift And Tired Light

This will describe the three additional contributions to red shifts – which are independent of velocity

Reexamination of the assumption that the red shift and the Hubble constant represent the velocity of remote galaxies leads to the realization that the Hubble constant relating velocity to distance is only based upon the questionable assumption that only velocity and the Doppler effect can cause red shift. This neglects any other contribution to the red shift such as the gravitational drag on traveling photons by interstellar gas and dust. The usual assumption relating red shift to velocity, (neglecting other contributions at large distances), may be a serious error and suggested the wrong theory of the expanding universe, along with the wrong theory of the acceleration of the expansion of the universe. It also initiated the wrong model of the big bang based upon the supposed expansion and the expansion acceleration, and the Dark Energy to power the supposed acceleration of the expansion.

As a surprising consequence of this extension of the theory of gravity and the associated equations, the important "tired light" suggestion of Zwicky is predicted and explained.

In addition to the Doppler shift contribution to the red shift, there are three additional contributions that are only due to gravity. They are caused by the loss of photon energy in leaving a massive body like the sun or for Quasars possibly containing a black hole. Even more important is loss of photon energy when traveling against gravity from the source of the photons. Also important is the linear dependence of the red shift on travel distance, called tired light, due to gravitational drag on dust and gas in interstellar space (without blurring, scattering collisions, or absorption and reemission).

At closer distances as in our milky way there are blue shifts as well as red shifts because of reduced red shift related to shorter travel distance

Red shift due to gravitational drag on gas and dust leaves behind some photon energy in the form of remaining motion of the encountered mass

Hubble Constant

We have also shown that the red shift and Hubble's constant need reexamination and cannot be used to determine star receding velocity and expansion of the universe because of the additional effect of three gravity interactions reducing photon energy.

This problem associated with the Hubble constant started in the 1930s with the excellent work of Hubble who determined through direct observations that the distance to stars with standard light output was linearly correlated with direct observations of the associated red shifts (increase of wavelength associated with reduction of photon energy). At first Hubble ascribed the red shift to an "…apparent Doppler effect…" and later voiced some doubts.

Others, including Einstein, believed that the red shifts showed that the universe is expanding although there were no direct observations of the receding motion of stars or galaxies.

We believe that the red shift and Hubble constant cannot be used to determine star velocity, and in certain cases even cannot be used to determine Quasar or star distance when black hole gravity significantly increases the red shift from Quasars.

Age Of Universe

The determination of the age of the universe, about 13.7 billion years, is based upon the value of the Hubble constant – which depends upon the assumption of the Doppler shift associated with the velocity as the major cause. This needs reevaluation of the assumptions because there are no direct confirming observations of velocity.

Dark Energy

The Hubble linear relationship between the red shift and the distances must become nonlinear when extrapolated for very for very remote galaxies. (Probably it is a log function of distance based upon my analysis of gravity and the mystery of Dark Matter.)

This problem with this use of the Hubble relationship occurs when the Hubble relationship is linearly extrapolated to very remote galaxies using the supernovae type Ia standard candle, the energy of the red shift photons at some extrapolated point must go through zero and then become negative (not reasonable) for very large distances. This is due to the loss of photon energy (longer wavelength) associated with the red shift and distance. Why can this extrapolation of the Hubble relationship be used to give an unlikely result?

If the deviation of the red shift from linearity at large distances is not recognized or appreciated then one reaches the speculative conclusion of Saul Perlmutter and others concerning increasing velocities and the apparent acceleration of remote galaxies – resulting in the questionable and unproven speculation about the need for Dark Energy.

A recent article on Dark Energy as "most profound problem" in physics illustrates a problem with the standard model of the universe – which is apparently accepted as the true mode – in spite of many other problems.

Saul Perlmutter (Perlmutter, 2003) and Adam Riess observed that very remote supernovae, type Ia (standard candles) were much further than predicted by their associated red shifts. To explain this, it was suggested that there was a source of Dark Energy that made these remote standard candles move faster than can be explained by the presumed expansion of the universe.

Actually there are other causes of red shifts – one for example is the well known gravitational red shift observed for photons emerging from stars.

Also, if we acknowledge that the observed red shifts do not prove that the universe is expanding, the problem of Dark Energy is extinguished – and cosmologists can proceed to more productive research.

Quasars

Many of the observations reported for quasars can be clarified. When the distance to a quasar is determined from the red shift and is much too large because of the gravitational effect of a massive black body in the quasar, the computed energy output will be incorrectly said to be surprisingly large.

In addition, when the transverse velocity of such a quasar is computed from the observed transverse angular motion, the computed transverse velocity based upon the very large distance (wrong) determined from the red shift is larger than C – and violates Einstein's limit.

Also when galaxies connected by stars streaming from one to another have their red shifts compared (Arp 1987, 1988), the presence of a massive black hole in one (contributing to the red shift by the gravitational effect) will make it appear much remote than the other – in conflict with the observed connection of the two galaxies by streaming stars. They are connected but with different red shifts due to different masses and probably a black hole in the apparently further one.

Why are there gaps in the distribution of observed red shifts? Probably there are gaps in the distribution of red shifts because of the gaps in the spacing of galaxies, which will cause gaps in the travel distances of photons to the observer.

Formation Of Universe

Note that with this theory of additional gravity, TAG, Newton's laws and Einstein's general relativity are still valid in our solar system where the distance, r, is small compared to galactic distances.

It can predict that computer modeling of the time for formation of stars, galaxies, and galactic structure using

Newton's constant will not agree with other determinations of the time, because of not using the additional gravity.

The observed gravitational lenses are also predicted without needing Dark Matter, just additional gravity from black holes or massive galaxies.

Before speculating on the early beginning of the universe, and the end of the universe, we should first try for a better understanding of the current universe.

An explanation of the apparent acceleration of the expansion of the universe has required the existence of the cosmological constant of Einstein, or of Dark Energy.

Why do stars form in strings, walls and with the observed webs of stars and galaxies?

Why is there a problem when the calculated time required for formation of cosmic structures such as strings and voids show a problem when compared to the deduced age of the universe?

Cosmic Microwave Background

We can predict and explain the Cosmic Microwave Background, CMB, without needing belief in the big bang.

There were prior, independent calculations of the low temperature of about 2.7 K in interstellar space, based upon calculations using the observed light from the stars and galaxies.

Black Holes

Another confusion in the understanding of the universe is connected with Black holes. We consider: the need for preservation of information, various ways of converting information, evaporation of mass and information in the form of photons, enabled by very high energy in the tail of the Maxwell Boltzmann energy distribution. The residue of the evaporating black hole is in the remaining form of a neutron star. The concept of other universes as an explanation for conservation of information appears to be pure speculation.

Singularities in explanations may be evidence of failure of theory.

Finally

Unfortunately the wrong models of the universe are embedded in the present generations of scientists (like in past centuries) and we will have to wait for the next generations to arrive at a true, simplified model of the universe. All I can do is share my insights in forms that may be available to them in the future.

Science including cosmology depends upon observations to suggest theories and even more important, on observations used to validate the predictions of theories.

The introduction of a new understanding of the true meaning of additional gravity is able to predict and explain many of the puzzling observations and mysteries of the universe.

According to the razor of William of Ockham, (check with Google search for our apparently correct spelling) the simplest explanation is preferred. If the new simple hypothesis is consistent with past observations and with future predictions, it should be preferable to the currently accepted model of the universe (that has many unexplained aspects).

A careful, open minded review of the points raised here could be beneficial to the future progress of the scientific community although the results could be upsetting to many intelligent and productive experts working in this field.

Many other surprising predictions and verifications of my hypothesized addition to the gravitational theory should be reviewed and possibly extended or critiqued by others in the field of astrophysics. If basically correct, the new hypothesis could result in a new, simplified, and more accurate view of the universe and could help future work of those in the field. If my new model of additional gravity is verified independently and accepted, many in the field of astrophysics may find their work easier and more productive.

Before speculating on the early beginning of the universe, and the end of the universe, we should first try for a better understanding of the current universe.

We should expect skepticism for a number of reasons including resistance to change. Under peer review by experts with other ideas, advances are sometimes in conflict with the common beliefs, and are hard to accept.

Examples include the past belief that the earth is flat, and that the universe revolved around the earth.

A serious, non-scientific problem is due to the Rice bowl effect (loss of funding). New and truer ideas could harm the funding, interests, and reputations of strong supporters of common beliefs.

There is a statement that progress advances by funerals after funerals – as new generations appear with their more open minds

People who know the "truth" and reject new ideas frequently hamper progress in science. This was the case in the time of Copernicus, and later in the time of Galileo. Fortunately, rejection is less dangerous now, particularly if external financial support is not needed.

Suggestion (with a smile)

In modern and past times a barrier to progress exists in the form of common knowledge and beliefs and also committees of known experts and review committees who vote on the acceptance of new ideas.

I suggest that we can save time and money by avoiding direct observations and only using committees of experts to rule on the truth of new ideas presented for acceptance or funding. There then will be no need to perform expensive experiments or make expensive observations. (Smile – just kidding, I hope.)

I am a scientist and strongly support research because the growth and survival of all depend upon scientific progress. Even science based upon wrong premises will have benefits in training future scientists, and can also result in related knowledge. An example is the growth and application of semiconductors and our computer industry and the Internet that were a byproduct of support by much of the other scientific research.

Advice

Scientists and engineers are a major resource for humanity, particularly if their activities are directed properly and productively, and if the necessary funding and freedom are provided.

The business leaders can have the necessary influence on the government to take the proper actions to help the country - and their own businesses can also benefit.

In any event, the future welfare of their children, grandchildren, relatives, and friends should also be an incentive for their activities to help the progress of science and the country.

Why this book

This is presented as one of my contributions to society, and probably can be of interest and value to future generations of scientists and others who are interested in, and want, or need a better understanding of the universe.

I expect to see objections from many who have worked in the field and who may have reputations as experts. However, if the objections are on an emotional basis, and without true support by observations, I will take this as another indication of the need for my current contribution that is designed to help future workers in the true understanding of the universe.

Speculation about the beginning of the universe and the ultimate end of the universe is interesting for others but I believe has no practical value to humanity because there is no way of influencing the beginning or end.

Actually humanity as we know it probably will cease to exist long before the end of the universe.

However, it is important to correctly understand the universe in its current state, as it exists in the present, so that knowledge can continue to help humanity survive and grow.

About the author

To help establish my scientific credentials, this will provide a very brief summary of the education, experience, and activities of the author, Sol Aisenberg, PhD. More details, including some publications, reports, and presentations are provided in Appendices.

Education

Sol Aisenberg holds a PhD in physics from MIT He graduated Cum Laude from Brooklyn College with a B.S. in Physics after graduating from Brooklyn Technical High School. He also has held part time appointments as a staff member at MIT, in the Physics Department, and the Research Laboratory of Electronics. When he had time, he was also a lecturer at Harvard Medical School, and a Visiting Research Professor in the Bioengineering Department of Boston University. He was elected as a member of Phi Beta Kappa, Sigma Xi (science), and Pi Mu Epsilon (math) honor societies.

Business Experience

He has held long time positions as Principal Investigator on contracts from many agencies. Included was line responsibility as General Manager, Principal Investigator, and Division President of the Space Sciences Division, a high technology division of the conglomerate, Whittaker Corporation, and as Division President and Principal Investigator of a high technology division, Applied Science Division of the conglomerate, Gulf+Western. These two were excellent conglomerates and left us alone to be self-funding through contracts.

Technology And Science

Dr. Aisenberg is a professional consultant, inventor, and scientist. For relaxation and as a Physicist, Aisenberg devotes a portion of his time in studying the puzzles, speculations, and misunderstandings in the standard model of the universe. He has identified one problem as the assumption that Newton's theory of gravity is supposed to also be universal and valid outside our solar system. Another insight is the understanding that the observed remote red shifts observed by Hubble only depends upon distance and is not due to the Doppler effect. These have led him to the solution of the shortcomings in the common belief in the standard model of the universe.

Aisenberg has over 19 years of experience in obtaining and running programs from many government agencies including NASA, NASA Headquarters, NASA Ames, NIH, WPAFB, HAFB, DARPA, ONR, and AFOSR, as well as programs from a medical foundation. Many of these programs were renewed for multiple years. As a result he has gained experience, capability, and additional knowledge in many disciplines.

Aisenberg is a generalist and has many publications and patents in many areas of activity and fields, such as instrumentation, lasers, medical devices, electronics, electro-optics, plasmas, ultra high vacuum, physical electronics, optics, artificial diamond material, bio-engineering, bio-technology, energy conversion systems, special materials, computers, video, software development, special application software, and financial analysis.

He has particular interest and experience in instrumentation and associated applications. Some of his work involved electrometer circuits for measuring ultra-high vacuum, electric arc research, plasmas, magneto-hydrodynamics, thin film deposition and characterization, Langumire probe plasma measurements, paramagnetic resonance, solar energy converters, gas lasers, electrode effects, electric arcs, insulating films, plasma accelerators, ion beam deposition, blood flow measurement, implanted glucose sensors, artificial hearts, biocompatibility, fuel cells, implanted oxygen sensors, biocompatible materials, optical fibers, stereo display, superconductors, diamond thin films, blood pressure measurement, microwave devices, medical devices, fast hand dryers, toys, eye view tracking, the effect of color in Dyslexia, and in instrumentation development.

He is a pioneer in the development and demonstration of diamond-like carbon, and introduced this name (DLC) to describe thin films of diamond produced by argon and carbon ion assisted deposition – these films are really pure carbon without hydrogen.

Affiliations And Accomplishments:

Dr. Aisenberg has over 134 publications, presentations, and reports plus a number of invited talks. Included are 17 dealing with electro-optics, 40 with instrumentation and medical devices, 91 involving thin films and plasmas, and 42 dealing with artificial diamond material, diamond-like carbon and related thin film technology. Also included are

8 awards, plus over 26 issued U.S. patents and various patents for clients with more patents pending.

Dr. Aisenberg has presented invited papers in the fields of diamond-like carbon, lasers, energy conversion, plasmas, thin films, medical devices, instrumentation, consulting, inventing, and superconductors.

He has seven IR-100 awards for important new products and an award for research in Helium-Neon Lasers. He was a reviewer for a number of professional and technical journals, and has reviewed proposals for the National Institute of Health and for the National Science Foundation.

He is a past member of Board of Directors of the Society of Professional Consultants. He served as a Vice-chairman of the IEEE Boston Consultants Network and also on the advisory group of the IEEE Boston Entrepreneurs Network. He has presented invited lectures on diamond thin films, on inventing, on consulting, and on technology transfer.

Aisenberg has been listed in editions of American Men of Science, Who's Who in the World, Who's Who in Finance and Industry, and Who's Who in Technology Today, among others.

Aisenberg is a current or former member of many professional societies including the American Physical Society, Division of Astrophysics, Division of Biological Physics, Division of Material Physics, Topical Group of Gravitation, the APS Vacuum Society, the Materials Research Society, the IEEE, Society of Photo-Optical Instrumentation Engineers (SPIE), and Association for Research in Vision and Optics (ARVO). He is a former member of technology transfer organizations, including the

Licensing Executives Society, (LES), and the Technology
Transfer Society, (TTS).

Sol Aisenberg, PhD
Tel: 508/651-0140
solaisenberg@comcast.net
saisenberg@alum.mit.edu

References And Reading Material

This represents some of the material located and read in the process of my analysis. Some are now in my library. The books were purchased through the Internet and from sources that sell copies of books out of print. The information available on the Internet also was very useful.

Anderson, J.D., Laing, P., Lau, P.A., Liu, E.L., Liu, A.S., Nieto, M.M., and Turyshev, S.G., *Indication, from Pioneer 10/11, Galileo, and Ulysses Data, of an Apparent Anomalous, Weak, Long-Range Acceleration*, Phys. Rev. Lett. 81, 14 (1998) pp. 2858-2861

Anderson,,J.D. Turyshev, S.G., and Nieto, M.M., *A Mission to Test the Pioneer Anomaly*, International Journal of Modern Physics, D. Vol. 11, No. 10 (2002) 1545-1551

Arp, H., *Seeing Red: Redshifts, Cosmology and Academic Science*, (Apeirion, Montreal, Canada, 1998). Aslso http://redshift.vif.com

Arp, H., *Quasars, Redshifts, and controversies*, (Interstellar Media, 1987).

Assis, A.K.T. and Neves, M.C.D., *History of 2.7 K Temperature Prior to Penzias and Wilson*, APEIRON, VOL. 2 July 1995, pp/ 79-84

Bahcall, N.A., *Large Scale Structure in the Universe"*, in *Unsolved Problems in Astrophysics*, edited by J. N.

Bahcall, and J. P. Ostriker, (Princeton University Press, NJ, 1997), pp. 61-91.

Bothun,G., *Modern Cosmological Observations and Problems*, (Taylor & Francis, London, 1998). Other modifications of Newton's law have been proposed along with discussions of the many problems in the current cosmological models.

Ferguson, K.. *Measuring the Universe*, (Walker and Company, N.Y., 1990).

Goldsmith, D., *The Astronomers*, (St. Martins Press, NY, 1991), pp. 36-44.

Gregory, J. *Fred Hoyle's Universe*, (University Press, Oxford, 2005)

Guth. A., *The Inflationary Universe*, (Perseus Books, Cambridge Massachusetts, 1997).

Harrison, E.R., *Cosmology: the Science of the Universe* (Cambridge University Press, Cambridge, 1981), p. 240, discusses the tired light concept of Zwicky.

Hawking,, S.W. *A Brief History of Time*, (Bantam Books, N. Y., 1998)

Hirschfield, AW., *Parallax*, (W. H. Freeman, N.Y., 2001).

F. Hoyle, F., Burbidge, G., and Narlikar, J.V., *A Different Approach to Cosmology*, (Cambridge University Press, N.Y., 2000).

Hubble, E., *A Relation between Distance and Radial Velocity among Extra Galactic Nebulae*, Proceedings of the National Academy of Science, vol.15, pp. 168-73 (1929).

Hubble, E., *The Observational Approach to Cosmology*, (Clarendon Press, Oxford England, 1937) p. 68.

Hubble, E., and Humason, M., ApJ. 74, 43 (1931)

Hubble, E., *The Realm of the Nebulae*, (Yale University, New Haven 1936, 1982).

Kragh, H., *Cosmology and Controversy*, (Princeton University Press, Princeton, N.J. 1996).

Kushner, R.P., *Extravagant Universe*, (Princeton University, NJ, 2002).

Lerner, E.J., *The Big Bang Never Happened*, (Vintage Book, N.Y. c 1992).

Lightman, A., *The Discoveries*, (Pantheon Books, N.Y., 2005)

Livo, M., *The accelerating universe*. (John Wiley, 2000)

Milgrom, M., arXiv: astro-ph/9810302 v1 20 Oct 1998.

Milgrom, M., ApJ. Vol. 270, pp. 365-370. (1983)

Milgrom, M., *Does Dark Matter Really Exist?* Scientific American, pp 42-52, (2002).

Narlikar, J.V., *Introduction to Cosmology*, (Cambridge University Press, Second Edition, 1993)

Peebles, P.J.E, *Principles of Physical Cosmology*, (Princeton University Press, N.J. 1993).

Perlmutter, S., *Supernovae, Dark Energy, and the Accelerating Universe*, Physics Today, April 2003, p. 53-60.

Perlmutter, S. http://supernova.lbl.gov/PhysicsTodayArticle.pdf

Rubin, V., Burstein, D, Ford, Jr.,W. K., and Thonnard, N., *Rotation Velocities of 16 Sa Galaxies and a Comparison of Sa, Sb, and Sc Rotation Properties*, Astrophys. J. 289: 81 (1985)

Rubin, V. and Ford, W. K., *Rotation of the Andromeda Nebula from a Spectroscopic Survey of Emission Regions*, Astrophysical Journal 159:379 (1970).

Saslaw, W.C.Z.., *Gravitational physics of stellar and galactic matter*, Cambridge University Press, Cambridge, 1985

Silk, J., *The Big Bang*, 3rd. ed., New York, (W. H. Freedman and Company, 2001)

Smolin, L., *The life of the Cosmos*, (Fordham University, N. Y., 1997)

Sofue, Y. and Rubin, V., *Rotation Curves of Spiral Galaxies*, in Annu. Rev. Astrophys. 2001. 39:137-74. Also http://

www.physics.ucla.edu/~cwp/articles/rubindm/rubindm.html

Spergel, D., *Dark Matter, in Unsolved Problems in Astrophysics*, edited by J. M. Bahcall and J.P. Postriker, (Princeton University Press, Princeton, N.J., 1997) pp.221-240.

Tifft, G. and Cocke, W. J., *Quantized Galaxy Red shifts,* University of Arizona, Sky & Telescope Magazine, Jan. 1987, pgs.19-21

Weinberg, S., *The First Three Minutes*, second edition, (Basic Books, New York, 1988)

Zwicky, F., *Redshift of Spectral Line*, Proc. Nat. Acad. Sci., 1929, vol. 15, pp. 773-9

Zwicky, F., *On the red shift of spectral lines through interstellar space*, PNAS 15:773-779, 1929

Zwicky, F., *On the Possibilities of a Gravitational Drag of Light*, Phys. Rev. Letters 34:Dec. 28, 1929

Extra reading

THIS ONE IS STRONGLY RECOMMENDED

An interesting collection of information about the many open questions in physics is available at www.openquestions.com/oq-cosmo.htm

Read it and you will be surprised at the degree of speculation and the admitted lack of understanding by many experts.

Methods of gravity determination using laboratory techniques and equipment are discussed at http://mist.npl.washington.edu/eotwash/gconst.html

Evidence for the Big Bang - Remote Sensing Tutorial http://rst.gsfc.nasa.gov/Sect20/A9.html

Techniques for measuring distances – describes 26 methods. http://www.astro.ucla.edu/~wright/distance.htm

Appendix A

About me, to help establish creditability

I am providing information about the misunderstood universe that I obtained during about ten years of my part time, obsessive review of the observation and theories reported by others.

The results surprised me. Even more, I was concerned about how many smart, educated, experts in the field could have been so misled. Apparently, as in past centuries, it is based upon the need to follow others, and not to upset the current beliefs.

While not an expert in cosmology, I am experienced as a Physicist in many fields, and this provided the ability to contribute in a new field, cosmology. There was an advantage because I had no fixed beliefs that might need to be overturned.

In order to help the reader judge my scientific credentials, the following appendixes, A1, A2, and A3 will provide documentation of my education experience, and accomplishments, including a list of my publications, reports, and presentations. This includes a list of my patents, along with a copy of my curriculum vitae.

Appendix A1

Brief Summary Of Experience, Background, And Activities Of Sol Aisenberg, PhD

Taken from my web site http://inventing-solutions.com in about 1994,

Education

Sol Aisenberg holds a PHD in physics from MIT He graduated Cum Laude from Brooklyn College with a B.S. in Physics after graduating from Brooklyn Technical High School. He also has held part time appointments as a staff member at M.I.T, in the Physics Department, and Research Laboratory of Electronics. When he had time, he was also a lecturer at Harvard Medical School, and a Visiting Research Professor in the Bioengineering Department of Boston University. He was elected as a member of Phi Beta Kappa, Sigma Xi (science), and Pi Mu Epsilon (math) honor societies.

Business Experience

He has held long time positions as Principal Investigator on contracts from many agencies. Included was line responsibility as General Manager, Principal Investigator, and Division President of a high technology division of Whittaker Corporation, and as President and Principal Investigator of a high technology division of Gulf+Western. From experience as a manager and officer in several large and small companies, he is aware of the need for practical results and bottom line results. His is an experimental and applied physicist with experience in Physical Electronics, Electromagnetic theory, Materials, and medical technology, and is a generalist with experience and results in many disciplines.

He is President and founder of the International Technology Group (ITG), a Technology Development and Information Company, and the Data Associates Division, which have been in operation, including consulting, since 1987.

Technology And Science

Dr. Aisenberg is a professional consultant, inventor, and scientist. He makes inventions that are practical, which can result in strong patents, can meet real market needs, and that use the lowest suitable technology. He uses both high and low technology solutions.

He is also involved in evaluating and improving the merits, worth, and validity of Intellectual Property, technology, and patents for clients and for his own inventions. He helps clients by finding legal patent loopholes, by evaluating and improving the strength of patent claims, and also helping in bypassing blocking patents.

Aisenberg consulted for over a year as senior advisor on Intellectual Property for major financial information company (Thompson Financial) and a global financial transaction company (Depository Trust Company). He also has consulted for law firms in Washington, DC, in New York, and in Boston on patent matters. As a result

Aisenberg is able to identify the technology that (a) can be patented, (b) how to frequently bypass competitive patents, and (c) how to make his patents and client patents bulletproof.

Currently, Aisenberg is preparing his patent applications designed to help solve the problems of energy shortages, in new medical devices, and in improving security. His approach is to find simple solutions – the simplest solutions generate the most profit.

He locates and works with independent agents who will help in licensing his inventions, and who will identify and qualify clients who realistically want new products devised for them. These agents will share in the resulting royalties for up to 20 years.

For relaxation and as a Physicist, Aisenberg devotes a portion of his time in studying the puzzles, speculations, and misunderstandings in the standard model of the universe. He has identified one problem as the assumption that Newton's theory of gravity is supposed to also be valid outside our solar system. Another insight is the understanding that the observed remote red shifts observed by Hubble only depends upon distance and is not due to the Doppler effect. These have led him to the solution of the shortcomings in the establishment belief in the standard model of the universe. A book on this subject is being prepared.

Experience

Aisenberg has over 19 years of experience in obtaining and running programs from many government agencies including NASA, NASA Headquarters, NASA Ames,

NIH, WPAFB, HAFB, DARPA, ONR, and AFOSR, as well as programs from a medical foundation. Many of these programs were renewed for multiple years. As a result he has gained experience, capability, and additional knowledge in many disciplines.

Technology

Aisenberg is a generalist and has many publications and patents in many areas of activity, such as instrumentation, lasers, medical devices, electronics, electro-optics, plasmas, ultra high vacuum, physical electronics, optics, artificial diamond material, bio-engineering, bio-technology, energy systems, special materials, computers, video, software development, special application software, and financial analysis.

He has particular interest and experience in instrumentation and associated applications. Some of his work involved electrometer circuits for measuring ultra-high vacuum, electric arc research, magneto-hydrodynamics, thin film deposition and characterization, Langumire probe measurements, paramagnetic resonance, solar energy converters, gas lasers, electrode effects, electric arcs, insulating films, plasma accelerators, ion beam deposition, blood flow measurement, implanted glucose sensors, artificial hearts, fuel cells, implanted oxygen sensors, biocompatible materials, optical fibers, stereo display, superconductors, diamond films, blood pressure measurement, microwave devices, medical devices, fast hand dryers, toys, and instrumentation.

He is a pioneer in the development and demonstration of diamond-like carbon, and introduced this name (DLC) to describe thin films of diamond produced by argon and carbon ion assisted deposition – these films are really pure carbon without hydrogen.

Aisenberg consults for international and domestic companies whose technologies and products he feels are unusually worthwhile. In this consulting he acts as a technology resource, business advisor, and practical problem solver.

Affiliations And Accomplishments:

Dr. Aisenberg has over 134 publications, presentations, and reports plus a number of invited talks. He has over 23 U.S. patents issued, and many others pending. He has seven IR-100 awards for important new products and an award for research in Helium-Neon Lasers. He was a reviewer for a number of technical journals, and has reviewed proposals for the National Institute of Health and for the National Science Foundation.

He is a past member of Board of Directors of the Society of Professional Consultants. He served as Vice-chairman of the IEEE Boston Consultants Network and also on the advisory group of the IEEE Boston Entrepreneurs Network. He has presented invited lectures on diamond thin films, on inventing, on consulting, and on technology transfer.

Aisenberg has been listed in editions of American Men of Science, Who's Who in the World, Who's Who in Finance and Industry, and Who's Who in Technology Today, among others.

Aisenberg is a current or former member of many professional societies including the American Physical Society, Division of Astrophysics, Division of Biological

Physics, Division of Material Physics, Topical Group of Gravitation, the APS Vacuum Society, the Materials Research Society, the IEEE, Society of Photo-Optical Instrumentation Engineers (SPIE), and Association for Research in Vision and Optics (ARVO). He is a former member of technology transfer organizations, including the Licensing Executives Society, (LES), and the Technology Transfer Society, (TTS).

Sol Aisenberg, PhD
Tel: 508/651-0140
E-mail: solaisenberg@comcast.net
http://inventing-solutions.com

Appendix A2

Publications, Reports, Presentations, and Patents of S. Aisenberg, PhD

1. S. Aisenberg, "Vacuum Studies," Report on Twelfth Annual Conference on Physical Electronics, MIT, Cambridge, MA (1952).

2. S. Aisenberg, "Ionization Gauge Control Circuit," Report on Thirteenth Annual Conference on Physical Electronics, MIT, Cambridge, MA (1953).

3. S. Aisenberg, "Ion Gauge Control Circuit and Ion Gauge Non-linearities," Report on Fifteenth Annual Conference on Physical Electronics, MIT, Cambridge, MA (1955).

4. S. Aisenberg, "Harmonic Analysis of Distribution Functions," Report on Sixteenth Annual Conference on Physical Electronics, MIT, Cambridge, MA (1956).

5. S. Aisenberg, "Probe Measurements in a Low Pressure Mercury Arc Plasma," Report on Seventeenth Annual Conference on Physical Electronics, MIT, Cambridge, MA (1957).

6. S. Aisenberg, "Ion Generation, Electron Energy Distributions, and Probe Measurements in a Low Pressure Mercury Arc," PHD Thesis in Physics Department, MIT, Cambridge, MA (1957) (unpublished).

7. S. Aisenberg, H. Statz, G. F. Koster, "Test of Spin Hamiltonian for Ion (3+) in Strontium Titanate," Phys. Rev. 116, 811 (1959)

8. S. Aisenberg, "Basic Processes in the Failure of High Power Microwave Windows," Raytheon Company Research Division, Technical Memo T-187 (August, 1959)

9. S. Aisenberg, "Theoretical Performance of Photoemissive Solar Energy Converters," Report on Twentieth Annual Conference on Physical Electronics, MIT, Cambridge, MA (1960)

10. S. Aisenberg, "Direct Measurement of Work Function Changes," Raytheon Company Research Division, Technical Memo T-231 (July, 1960)

11. S. Aisenberg, "Theoretical Performance of Photo emissive Solar Energy Converters," Invited paper given at meeting of New England Section of American Chemical Society, Boston, MA (October, 1960)

12. S. Aisenberg, "Theoretical Performance of Photo emissive Solar Energy Converters," Raytheon Company Research Division, Technical Report R-59 (April, 1961)

13. S. Aisenberg, "Modern Probe Techniques for Plasma Diagnosis," Invited Paper presented at the Third Annual Symposium on the Engineering Aspects of Magneto hydrodynamics (at the University of Rochester, March 1962), Engineering Aspects of Magneto hydrodynamics, Third Annual Symposium, (Gordon and Breach, 1964)

14. S. Aisenberg, "Modern Probe Techniques for Plasma Diagnosis," Invited talk presented at MIT Microwave Laboratory Seminar on Microwave and Plasmas, MIT, Cambridge, MA (April, 1962)

15. S. Aisenberg, "Multiple Probe Measurements in High Frequency Plasmas," Proceedings of the 23rd Annual Conference on Physical Electronics, MIT, Cambridge, MA (1963)

16. S. Aisenberg, "The Effect of Helium on Electron Temperature and Electron Density in Rare Gas Lasers," Appl. Phys. Lett. 2, 187 (1963)

17. S. Aisenberg, "Measurement of Plasma Properties by Means of Probe Techniques," Invited paper presented at the National Electronics Conference, Chicago (October, 1963). Proceedings of the National Electronics Conference 19, 668 (1963)

18. S. Aisenberg, "The Sticking Coefficient, the Optical Transmission, and the Oxidation of Thin Metallic Films," Proceedings of the 10th National Symposium, American Vacuum Society, Boston, MA (October, 1963)

19. S. Aisenberg, "Multiple Probe Measurements in Rare Gas Plasmas," J. Appl. Phys. 35, 130 (1964)

20. S. Aisenberg, "Optical Gain and Absorption Measurements in Rare Gas Plasmas," Proceedings of the 24th Annual Conference on Physical Electronics, MIT, Cambridge, MA (1964)

21. S. Aisenberg, D. V. Missio, and P. A. Silberg,

"Performance of the Plasma Theta-Pinch for Laser Pumping," J. Appl. Phys. 35, 3625 (1964)

22. S. Aisenberg, "An Optical Study of Performance Limitations of Plasma Lasers," Invited paper presented at the National Electronics Conference, Chicago (October, 1964). Proceedings of the National Electronics Conference 20, (1964)

23. S. Aisenberg, "The Effect of Gap Spacing Upon High-Voltage Vacuum Breakdown," Proceedings of the 25th Annual Conference on Physical Electronics, MIT, Cambridge, MA (1965)

24. S. Aisenberg, P. Hu, V. Rohatgi, and S. Ziering, "A Study of Electrode Effects in Crossed Field Accelerators," Summary Report, Contract NASA-1014, prepared for National Aeronautics and Space Administration (1965).

25. S. B. Afshartous, S. Aisenberg, V. Rohatgi, and C. G. Smith, "Investigation of Cathode Phenomena in the Mercury Arc," Final Report, Contract No. AF30(602)-3093, prepared for Rome Air Development Center (1965).

26. S. Aisenberg and V. Rohatgi, "A Study of Arc Constriction Processes," Proceedings of the Seventh Symposium on the Engineering Aspects of Magneto hydrodynamics (March, 1966)

27. S. Aisenberg and V. Rohatgi, "Measured Tangential Electrode Forces for an Arc in a Transverse Magnetic Field," Appl. Phys. Lett. 8, 194 (1966)

28. S. Aisenberg and V. Rohatgi, "Composite Metallic and Dielectric Insulators for High Current Arc Electrodes," Rev. Sci. Instrum. 37, 1603 (1966)

29. V. Rohatgi and S. Aisenberg, "Tangential Momentum Transfer to the Electrodes of a Magnetically Accelerated Arc," Presented at the American Physical Society Meeting, Boston, MA (1966). Bull. Am. Phys. Soc. 12, 780 (1967)

30. S. Aisenberg and V. Rohatgi, "A Study of Electron Emission Processes at Arc Cathodes," Presented at the American Physical Society, Boston, MA (1966). Bull. Am. Phys. Soc. 12, 811 (1967)

31. S. Aisenberg, "Plasma Propagation Theory of Arc Retrograde Motion," Proceedings of the Eighth Symposium on the Engineering Aspects of MHD (March, 1967)

32. S. Aisenberg, "The Anomalous Dielectric Constants of Very Thin Insulating Films", Proceedings of the 27th Annual Conference on Physical Electronics, MIT, Cambridge, MA (1967). Bull. Am. Phys. Soc. 12, 984 (1967)

33. S. Aisenberg and S. E. Nydick, "Electron Attaching Additives for the Modification of Re-entry Plasmas," Scientific Report No. 1, AFCRL-67-0214, Contract No. AF19(628)-5097 (March, 1967)

34. S. Aisenberg, P. Hu, V. Rohatgi, and S. Ziering, "Plasma Boundary Interactions," NASA Contractor Report, NASA-CR-868 (August, 1967).

35. S. Aisenberg and V. Rohatgi, "Ion Drag and Current Partitioning at the Cathode of a Plasma Accelerator," Presented at the AIAA Joint Electric Propulsion and Plasma Dynamics Conference, Colorado. AIAA Paper No. 67-657 (September, 1967).

36. S. Aisenberg, "The Use of Chemical Additives for the Alleviation of the Plasma Sheath Problem," Presented as an invited paper at the Conference on Applications for Plasma Studies to Re- Entry Vehicle Communications, U.S. Air Force Avionics Laboratory, Dayton, OH (October 3-4, 1967)

37. S. Aisenberg, "A Study of the Use of Chemical Additives for the Alleviation of the Plasma Sheath Problem," Scientific Report No. 2, AFCRL-67-0693, Contract No. AF19(628)-5097 (December 1967)

38. S. Aisenberg and S. E. Nydick, "The Study of Plasma Surrounding Re-Entry Bodies and Resultant Interaction with Microwave Radiation," Final Report AFCRL-68-0239, Contract No. AF19(628)-5097 (1968)

39. S. Aisenberg, P. N. Hu, V. Rohatgi, and S. Ziering, "Plasma Boundary Interactions-II," NASA Contractor Report, NASA-CR-1072 (1968)

40. S. Aisenberg, D. Cipolle, J. Feather, J. Newman, and V. Rohatgi, "Absorptivity/Emissivity Measuring Systems," NASA Contract No. NAS8-20581 (1968).

41. S. Aisenberg and V. Rohatgi, "Study of the Deposition of Single Crystal Silicon, Silicon Dioxide and Silicon Nitride on Cold-Substrate Silicon," Interim Report,

prepared for NASA Electronics Research Center under Contract No. NASA12-541 (1968).

42. S. Aisenberg and R. Chabot, "Study of the Deposition of Single Crystal Silicon, Silicon Dioxide, and Silicon Nitride on Cold-Substrate Silicon," Final Report prepared for NASA Electronics Research Center under Contract No. NASA12-541 (1969).

43. S. Aisenberg, "Study of Plasma Boundary Physics," Presented at NASA Plasma Boundary Layer Research Conference at the NASA/Langley Research Center, Hampton, VA (April 10, 1968).

44. V. Rohatgi and S. Aisenberg, "A System for the Measurement of the Absorptivity and Emissivity of a Vehicle Surface During Flight Conditions," Proceedings of the 23rd Annual Instrument Society of America Conference, New York (October, 1968)

45. S. Aisenberg, "The Deposition of Single Crystal Silicon Films on Cold Single Crystal Silicon Substrates," Proceedings of the 1968 Government Microcircuit Applications Conference (GOMAC), Gaithersburg, MD (October, 1968)

46. S. Aisenberg, "The Ion Beam Deposition of Single Crystal Silicon Films on Cold Single crystal Silicon Substrates," Proceedings of the 15th National Vacuum Symposium, Pittsburgh, PA (October, 1968)

47. S. Aisenberg, "Physical Processes For The Removal of Free Electrons in Re-Entry Type Plasma Sheaths," Bull. Amer. Phys. Soc. 13, 1494 (1968).

48. S. Aisenberg, P. N. Hu, and K. W. Chang, "Modification of the Plasma Sheath by Rapidly Evaporating Liquid Additives," Space Sciences Incorporated, Semi-Annual Report prepared for the Department of the Army, U.S. Army Research Office, Durham, North Carolina and sponsored by Advanced Research Projects Agency under Contract No. DAHCO4-68-C-0031 and ARPA Order No. 553 (April, 1969).

49. S. Aisenberg and V. Rohatgi, "Ion Drag and Current Partitioning at the Cathode of a Plasma Accelerator," AIAA Journal 7, 502 (March, 1969)

50. S. Aisenberg and V. Rohatgi, "Plasma Boundary Interactions-III," Prepared under Contract No. NASA-1703 (August, 1969)

51. S. Aisenberg and K. W. Chang, "An RF Coil System for the Measurement of Plasma Electrical Conductivity," Scientific Report No. 1, AFCRL-70-0033, prepared for Air Force Cambridge Research Laboratories (October, 1969)

52. S. Aisenberg and K. W. Chang, "A Study of the Effect of Ion Interactions in Plasma Accelerators," Sixth NASA Intercenter and Contractor Conference on Plasma Physics, Langley Research Center, Hampton, VA (December, 1969)

53. S. Aisenberg and R. W. Chabot, "An Ion Beam Deposition Technique for the Formation of Thin Films of Insulating Carbon," Presented at the Second Annual Symposium of American Vacuum Society and

the New England Section of the Surface Division of the American Vacuum Society, Waltham, MA (May, 1970)

54. S. Aisenberg and K. W. Chang, "High Conductivity Plasma Produced By Runaway Electrons in a Pulsed Electrodeless MHD Generator", Presented at 1970 American Physical Society Spring Meeting, Washington, D.C. Bull. Amer. Phys. Soc. 15, 588 (1970)

55. S. Aisenberg and K. W. Chang, "A Wide Range Probe for Measurement of Plasma Electrical Conductivity", Presented at 1970 American Physical Society Meeting, Washington, D.C., Bull. Amer. Phys. Soc. 15, 588 (1970).

56. S. Aisenberg, "A Catheter Coil System for the Measurement Of Blood Flow Velocity, Flow Channel Cross-Section and Flow Rate," Presented at the 7th Annual Rocky Mountain Bio-Engineering Symposium and the Eighth International ISA Biomedical Sciences Instrumentation Symposium, Denver, CO (May 4-6, 1970)

57. S. Aisenberg and R. W. Chabot, "Ion Beam Deposition of Thin Films of Insulating Carbon," Proceedings of the Government Microcircuit Applications Conference, (GOMAC), New Jersey (October 6-8, 1970)

58. S. Aisenberg and P. N. Hu, "The Removal of Free Electrons in a Thermal Plasma by Means of Rapidly Evaporating Liquid Additives," Presented at the

Fourth Plasma Sheath Symposium, Hampton, VA (October 13-15, 1970)

59. S. Aisenberg and K. W. Chang, "A Non-Protruding Conductivity Probe System for Re-Entry Plasma Diagnostics," Presented at the 4th Plasma Sheath Symposium, Hampton, VA (October 13-15, 1970)

60. S. Aisenberg and R. Chabot, "Ion Beam Deposition of Thin Films of Diamond-Like Carbon," Presented at the 17th National Vacuum Symposium, Washington, D.C. (October 20-23, 1970)

61. S. Aisenberg and K. W. Chang, "A Pulsed MHD Generator Design for a High-Conductivity, High-Efficiency, Electrodeless MHD AC Generator," Bull. Amer. Phys. Soc. 15, 1484 (1970)

62. S. Aisenberg and P. N. Hu, "A Theoretical and Experimental Study of the Basic Properties of Plasmas" Final Report, Contract No. F19628-68-C-0127, prepared for Air Force Cambridge Research Laboratories, AFCRL-71-0018 (November, 1970)

63. S. Aisenberg and R. Chabot, "Ion Beam Deposition of Thin Films of Diamond-Like Carbon," Invited paper presented to New England Chapter Thin Film Division, American Vacuum Society (December, 1970)

64. S. Aisenberg and R. Chabot, "Ion Beam Deposition of Thin Films of Diamond-Like Carbon," J. Vac. Sci. Technol. 8, 1 (January/February, 1971)

65. S. Aisenberg and P. N. Hu, "Modification of Plasma by

Rapidly Evaporating Liquid Additives," Final Report, Contract No. DAHCO4-68-C-0031, prepared for the Department of the Army, Sponsored by Advanced Research Projects Agency (February, 1971). FOR OFFICIAL USE ONLY

66. S. Aisenberg and R. Chabot, "Ion Beam Deposition of Thin Films of Diamond-Like Carbon," J. Appl. Phys. 42, 2953 (1971)

67. S. Aisenberg and K. W. Chang, "A Wide Range RF Coil System for the Measurement of Plasma Electrical Conductivity," Proc. IEEE 59, 710 (April, 1971)

68. S. Aisenberg, "Ion Beam Deposited Carbon Films," Invited Seminar presented at Pennsylvania State University, Department of Material Sciences (May, 1971).

69. S. Aisenberg and R. Chabot, "Deposition of Carbon Films with Diamond-Properties," Proceedings of the 10th Biennial Conference on Carbon. Lehigh University (July, 1971)

70. S. Aisenberg and K. W. Chang, "An Implantable Glucose Fuel Cell Sensor System," Final Report under Contract No. JDF-70-1003, Joslin Diabetes Foundation, Boston, MA (June 14, 1971)

71. S. Aisenberg, K. W. Chang, and J. S. Soeldner, "Some Bacteria and Fungus Inhibitors for In-Vitro Testing of an Implantable Glucose Sensor," (Whittaker Space Sciences Division Report)

72. S. Aisenberg and R. Chabot, "System for Mass

Screening for Lead Poisoning," Proceedings of the Seventh Annual Meeting of Association for the Advancement of Medical Instrumentation (April, 1972)

73. S. Aisenberg, K. W. Chang, and P. N. Hu, "Ion Velocity and Ion Microfield Emission in a Plasma", Final Report, Contract No. NAS2-1703, prepared for the National Aeronautics and Space Administration (June, 1972)

74. S. Aisenberg and K. W. Chang, "Chemical Additives and Diagnostics for High Temperature Air Plasmas," Final Report, Contract F19628-71-C-007, prepared for the Air Force Cambridge Research Laboratories (November, 1971)

75. K. W. Chang, S. Aisenberg, and J. S. Soeldner, "In-Vitro Tests of an Implantable Glucose Sensor," Proceedings of the 25th Conference on Engineering in Medicine and Biology, 58 (1972)

76. S. Aisenberg and R. W. Chabot, "Versatile Coating Systems for Ion Beam Deposition, Ion Plating, Sputter Deposition and Sputter etching," Presented at the 19th National Symposium of the American Vacuum Society (October, 1972). Bull. Amer. Phys. Soc.

77. S. Aisenberg and R. W. Chabot, "Physics of Ion Plating and Ion Beam Deposition," Presented at the 19th National Symposium of the American Vacuum Society (October, 1972). Bull. Amer. Phys. Soc.

78. S. Aisenberg and R. W. Chabot, "Physics of Ion

Plating and Ion Beam Deposition," J. Vac. Sci. Technol. 10(1), 104 (1973)

79. J. S. Soeldner, K. W. Chang, S. Aisenberg, and J. R. Hiebert, "Status Report on the Glucose Sensor Program," Joslin Diabetes Foundation, Inc., Boston, MA (October, 1972)

80. S. Aisenberg and K. W. Chang, "The Limitation of Magnetohydrodynamically Operated Artificial Heart Pumps," Proceedings of the Eighth Annual Meeting of the Association for the Advancement of Medical Instrumentation, Washington, D.C., Med. Instrumentation 7(1), 41 (1973)

81. K. W. Chang, R. W. Chabot, and S. Aisenberg, "A Fuel Cell Sensor for Alcoholic Breath Measurements," Presented at the Eighth Annual Meeting of the Association for the Advancement of Medical Instrumentation, Washington, D.C., Med. Instrumentation 7(1), 46 (1973)

82. S. Aisenberg and R. W. Chabot, "A Fluorometric System for Drug Detection," Presented at the Eighth Annual Meeting of the Association for the Advancement of Medical Instrumentation, Washington, D.C. (March 21-24, 1973). Med. Instrumentation, 7(1), 46 (1973)

83. K. W. Chang, S. Aisenberg, J. S. Soeldner, and J. M. Hiebert, "Validation and Bioengineering Aspects of an Implantable Glucose Sensor," Amer. Soc. Artif. Internal Organs 19, 352-385 (1973)

84. J. S. Soeldner, K. W. Chang, S. Aisenberg, and J. M. Hiebert, "Progress Towards an Implantable Glucose

Sensor and an Artificial Beta Cell," Temporal Aspects of Therapeutics, eds. J. Urquhart and F. E. Yates, (Plenum Press, New York-London, 1973), pp. 181-207

85. J. S. Soeldner, K. W. Chang, S. Aisenberg, J. M. Hiebert, and R. H. Egdahl, " Diabetes Mellitus. A Bioengineering Approach-An Implantable Glucose Sensor," Diabetes Mellitus, Chapter 20. Fogarty International Center Series on Preventive Medicine, Vol. 4, S. S. Fajans, ed. DHEW Publication No. (NIH) 76-854. Department of Health, Education and Welfare, Public Health Service, National Institutes of Health, pp. 267-277 (1976)

86. S. Aisenberg and K. W. Chang, "The Study of Improved Electrodes for Ion Lasers and for Molecular Gas Lasers," Final Report for Office of Naval Research, Contract No. N00014-71-0037, Space Sciences Division-P-677FR (November, 1972)

87. S. Aisenberg and R. W. Chabot, "Investigation of the Properties of Thin Insulating Films Deposited with an Ion Beam System." Final Report No. AFCRL-TR-73-0176 for Air Force Cambridge Research Laboratories under Contract No. F19628-72-C-0291 (January, 1973)

88. "Status Report II on the Glucose Sensor Program", with the Joslin Diabetes Foundation of Boston, MA (May, 1973)

89. "Status Report III on the Glucose Sensor Program", with the Joslin Diabetes Foundation of Boston, MA

(April 5, 1973)

90. J. S. Soeldner, J. M. Hiebert, K. W. Chang, and S. Aisenberg, "In Vitro and In Vivo Experience with a Miniature Glucose Sensor," Proceedings of the 1973 Annual Meeting of the American Diabetes Association, Chicago (1973)

91. J. S. Soeldner, J. M. Hiebert, K. W. Chang, and S. Aisenberg, "The Development of a Glucose Sensor Suitable for Implantation," Presented at the Eighth International Diabetes Congress, Brussels, Belgium (1973)

92. S. Aisenberg, K. W. Chang, and J. S. Soeldner, "The Artificial Beta Cell - An Implantable Biofeedback Controlled System for Administering Insulin," An invited paper presented at the Academy of Pharmaceutical Sciences Symposium, Boston, MA (1973)

93. K. W. Chang, S. Aisenberg, J. S. Soeldner, and J. M. Hiebert, "Basic Principle and Performance of a Viable Glucose Sensor Suitable for Implantation," Digest of the 10th International Conference on Medical and Biological Engineering, (The Conference Committee of the 10th International Conference on Medical and Biological Engineering, German Democratic Republic, Dresden,) pp. 56 (1973)

94. S. Aisenberg and R. W. Chabot, "Ion Beam Deposited Carbon Coatings for Bio-Compatible Materials," Final Report under Contract NIH-73-2919 for Division of Blood Diseases and Resources, National

Heart and Lung Institute, NIH (December, 1973)

95. J. S. Soeldner, K. W. Chang, S. Aisenberg, J. M. Hiebert, and R. H. Egdahl, "Progress Report - Artificial Implantable Beta- Cell," Proceedings of the 1973 International Conference on Cybernetics and Society (November 5-7, 1973). Institute of Electrical and Electronics Engineers, Inc., NY, NY

96. K. W. Chang, and S. Aisenberg, "Development of an Implantable Oxygen Sensor," Annual Report prepared for National Heart and Lung Institute, NIH on Contract No. N01-HL-3-3003R, Report No. Space Sciences Division-P-708AR#1 (June 1974)

97. S. Aisenberg and R. W. Chabot, "Ion Beam Deposited Carbon Coatings for Biocompatible Materials," Comprehensive Report prepared for National Heart and Lung Institute, NIH on Contract No. NIH-N01-HB-3-2919, Report No. Space Sciences Division-P-711CR (November, 1974)

98. K. W. Chang, and S. Aisenberg, "An Implantable Oxygen Sensor for Continuous Measurement of Blood Oxygen Tension," Med. Instrumentation, 9, No. 1, 74 (Jan/Feb, 1975)

99. S. Aisenberg, K. W. Chang, and J. S. Soeldner, "Response Time Mechanisms and Measurements for Implantable Glucose Sensors," Med. Instrumentation, 9, No. 1, 40 (Jan/Feb, 1975)

100. J. R. Guyton, K. W. Chang, S. Aisenberg, and J. S. Soeldner, "Activity of Various Endogenous Compounds and Pharmacologic Agents at a Glucose

Oxidizing Platinum Electrode," Med. Instrumentation, 9, No. 5 (Sept/Oct, 1975), pp. 227-232

101. S. Aisenberg, J. S. Soeldner, K. W. Chang, and C. K. Colten, "A Bioengineering Approach for the Treatment of Diabetes Mellitus," Topics in Diabetes Mellitus, (Wm. Heineman, London)

102. R. W. Chabot and S. Aisenberg, "Blood Compatibility of Ion Beam Deposited Carbon Coatings," Proceedings of the 28th Annual Conference on Engineering in Medicine and Biology (1975)

103. K. W. Chang, T. F. Kordis, S. Aisenberg, H. Hechtman, and R. Dennis, "Animal Trials of a Stable Intravascular Oxygen Sensor for Continuous Measurement of Blood CO2," Proceedings of the 28th Annual Conference on Engineering in Medicine and Biology (1975)

104. K. W. Chang and S. Aisenberg, "Development of An Implantable Oxygen Sensor: In Vivo Validation and Refinement," Annual Report for period June, 1974-75, prepared for Division of Lung Diseases, NIH, Report No. Space Sciences Division-P-708AR#2

105. R. W. Chabot and S. Aisenberg, "Continued Measurement and Study of the Blood Compatibility of Ion Beam Deposited Carbon: Investigation of Dominant Physical and Chemical Factors," Prepared for Division of Heart and Vascular Diseases, NIH, Final Report (1975)

106. D. A. Gough, S. Aisenberg, C. K. Colton, J. Giner, and J. S. Soeldner, "The Status of Electrochemical

Sensors for In Vivo Glucose Monitoring," Published in Blood Glucose Monitoring - Methodology and Clinical Application of Continuous In Vivo Glucose Analysis, ed. Prof. Dr. J. D. Druse-Jarres, (G. Thieme Verlag, Stuttgart, 1977)

107. D. A Gough, S. Aisenberg, C. K. Colton, J. Giner, and J. S. Soeldner, "The Status of Electrochemical Sensors for In Vivo Glucose Monitoring," Horm. Metab. Res. (1976)

108. P. R. Liebman, M. T. Pattern, R. D. Dennis, K. W. Chang, S. Aisenberg, and H. B. Hechtman, "Continuous Monitoring of In-Vivo Oxygen Tension with a Fuel Cell," Surgical Forum, 27 (1976)

109. S. Aisenberg and M. Stein, "Diamond-Like Carbon Films - Factors Leading to Improved Biocompatibility," Proceedings of the 13th Biennial Conference on Carbon (July 18-22, 1977), Irvine, CA. Extended Abstracts, pp. 87

110. S. Aisenberg and M. Stein, "Continued Studies of Ion Beam Deposited Carbon as a Blood Compatible Material," Annual Report, Space Sciences Division-P-836A for Contract NIH-NO1-HB-3-2919 National Institutes of Health (1977)

111. S. Aisenberg and M. Stein, "Ion Beam Deposited Carbon Films and Factors Important to Improved Biocompatibility," Proceedings of the AAMI 13th Annual Meeting, p. 6 (1978)

112. S. Aisenberg and M. Stein, "Development and Evaluation of an Implantable Oxygen Sensor,"

Proceedings of the AAMI 13th Annual Meeting, p. 7 (1978)

113. S. Aisenberg and M. Stein, "Factors Important in Applications of Biocompatible Materials," Extended Abstracts of the 14th Carbon Conference (1979)

114. M. Stein, and S. Aisenberg, "Evaluation of Ion Deposited Carbon Films," Extended Abstracts of the 14th Carbon Conference (1979)

115. S. Aisenberg and M. Stein, "Development and Evaluation of an Implantable Fuel Cell Oxygen Sensor." Final Report NIH No. HR- 33003-IF

116. M. Stein, S. Aisenberg, and J. M. Stevens, "Ion Plasma Deposition of Hermetic Coatings for Optical Fibers," Proceedings of the 82nd Annual Meeting of the American Ceramic Society (November, 1980)

117. S. Aisenberg and M. Stein, "The Use of Ion-Beam Deposited Diamond-Like Carbon for Improved Optical Elements for High Powered Lasers," Proceedings of the 12th Annual Symposium on Optical Materials for High Power Lasers (1980)

118. S. Aisenberg and M. Stein, "The Moisture Protection of Strong Optical Fibers," Interim Report, RADC-TR-80-252, Rome Air Development Center (1980)

119. S. Aisenberg, M. Stein, J. Stevens, and B. Bendow, "Ion Deposited Hermetic Coatings for Optical Fibers," Presented at the Fiber Optic Sensor Systems Workshop (FOSS) May, 1981

120. M. Stein, S. Aisenberg, and J. Stevens, "Ion-Plasma Deposition of Carbon-Indium Hermetic Coatings for Optical Fibers," Proceedings of the 1981 Conference on Lasers and Electro-Optics (CLEO) June, 1981

121. S. Aisenberg and M. Stein, "Novel Materials for Improved Optical Disk Lifetimes," Proceedings of the 15th International Technical Symposium of the Society of Photo and Electro-Optic Engineers, (SPIE), August, 1981

122. M. L. Stein, S. Aisenberg, B. Bendow, and BDM Corporation, "Studies of Diamond-Like Carbon Coatings for Protection of Optical Components," Proceedings of the 13th Annual Symposium on Optical Materials for High Power Lasers. November, 1981

123. S. Aisenberg, "Improved Hermetic Coatings for Optical Fibers." Radiation Curing VI Conference proceedings. September, 1982. Society of Manufacturing Engineers, Chapter 12, pp. 14-32

124. S. Aisenberg, "Properties and Applications of Diamond-Like Carbon Films." Proceedings of the American Vacuum Society, 30th National Symposium 1983, p. 132

125. S. Aisenberg and M. Stein, "The Moisture Protection of Optical Fibers," Final Report, Contract No. F19628-78-C-0180, RADC/ESM, Hanscom AFB. December, 1983

126. S. Aisenberg, "Improved Visual Display of Three-Dimensional Information." Final Report, NASA contract number NAS2-12083. September 1985

127. S. Aisenberg, "Technology of Diamondlike Carbon and its Applications," Invited lecture presented at a meeting of the American Association for Crystal Growth, Mid Atlantic Section, N. J., October 22, 1987

128. S. Aisenberg, "Early Applications of the Meissner Effect," First Conference on Superconductor Markets, Kessler Marketing Intelligence, Boston, December 15, 16, 1987

129. S. Aisenberg, "Novel Applications of the Meissner Effect to Space and Defense Problems," International Superconductor Applications Convention, Los Angeles, February 17, 1988

130. S. Aisenberg and F. M. Kimock, "Ion Beam and Ion-Assisted Deposition of Diamond-like Carbon Films," in Preparation and Characterization of Amorphous Carbon Films, Material Sciences Forum, Volumes 52 & 53 (1990), edited by John J. Pouch and Samuel A. Alterovitz (NASA Cleveland, Ohio) (Trans Tech Publications LTD, Aedermannsdorf, Switzerland, 1990)

131. Sol Aisenberg, "The Role of Ion Assisted Deposition in the Formation of Diamond-Like Films," J. Vac. Sci. Technol., A 8(3), 2150-2154 (1990)

132. Sol Aisenberg, Some Comments on Diamond-Like Carbon, Diamond, and Hard Carbon Materials Including Terminology, Deposition Processes, Composition, and Applications. Technical Note: TN-90-01, Applied Diamond Technology, 1990

133. Sol Aisenberg, Initial and Clarifying Publications Related to Diamond-Like and Hard Carbon. Technical Note: TN-90-02. Applied Diamond Technology, 1990

134. S. Aisenberg, Diamond-like Carbon Deposition Technology for Improved Barrier Films, Invited paper, Fourth International Conference on Vacuum Web Coating, Reno, October 31-November 2, 1990, (Bakish Materials Corporation, Englewood, N.J.)

135. S. Aisenberg, Some Problems and Advantages of International Technology Transfer, Proceedings of the International Congress on Technology and Technology Exchange, Champion Pennsylvania, November 6-8, 1990, p. 1

136. Sol Aisenberg, Practical Applications of Diamond thin films, Invited talk, New England Combined Chapter of the American Vacuum Society, March 20, 1991, Bedford, MA.

137. S. Aisenberg, A. Altshuler, and J. L. Sprague, Physical Deposition of Diamond - Technology, Properties, and Applications, Presented at the Second International Symposium on Diamond Materials, in the 179th Meeting of the Electrochemical Society, Washington, DC, May 5-10, 1991

138. S. Aisenberg, Locating and Assessing Available Technology, Presented at the Second Annual Technology Transfer Conference, Worcester Polytechnic Institute, Worcester, MA, May 29, 1991

139. Anatoly Altshuler and Sol Aisenberg, Low Temperature Deposition of Artificial Diamond Material by Means

of Halogen Chemical Transport Reaction, Presented at the 1991 Meeting of the Materials Research Society, Boston, MA, Abstract Book, p. 250

140. Sol Aisenberg, Too Much Technology: How to Determine the Suitability of Technology for Commercial Applications. Venture Newsletter, October 1991. WPI Venture Forum, Worcester Polytechnic Institute, Worcester, MA

141. Sol Aisenberg, Measurement and Analysis of Binocular Rapid Eye Movement, Including Saccades, Saccadic Prediction, fixations, Nystagmus, and Smooth Pursuit. Presented at the Annual Meeting of the Association for Research in Vision and Ophthalmology, Fort Lauderdale, FL, April 21-26, 1996

142. Sol Aisenberg , A Simplified Model of the Universe: Clarifying dark matter, dark energy, and the big bang, Poster, New England Section Fall Meeting 2004 October 22-23, 2004

143. Sol Aisenberg, Expanding Gravity, American Physical Society, New England Section Spring Meeting, MIT Cambridge, April 1-2, 2005.

144. Sol Aisenberg, Dark Matter and Dark Energy Discovered, Presented at the New England 2006 Spring Meeting of the American Physical Society, Boston, April 1, 2006.

# # #

Some Patents and Patent Disclosures with S. Aisenberg as Inventor or Co-inventor

1. Apparatus for Measurement of Plasma Conductivity,

U.S. Patent No. 3,525,931. August 25, 1970.

2. Magneto hydrodynamic Generator, U.S. Patent No. 3,660,700, May 2, 1972.

3. Extrusion System, U.S. Patent No. 3,675,451, July 11, 1972.

4. Doppler Shift System, U.S. Patent No. 3,795,448, March 5, 1974.

5. Blood Glucose Monitoring System, U.S. Patent No. 3,837,339, January 6, 1975.

6. Apparatus for Film Deposition, U.S. Patent No. 3,904,505, September 9, 1975.

7. Perforated Wall Hollow-Cathode Ion Laser, U.S. Patent No. 3,931,589, January 6, 1975.

8. Dual Field Excitation for a Carbon Dioxide Laser, U.S. Patent No. 3,943,465, March 9, 1976.

9. Film Deposition, U.S. Patent No. 3,961,103, June 1, 1976.

10. Apparatus for Measuring Alcohol Concentration, U.S. Patent No. 3,966,579, June 29, 1976.

11. Noise Rejecting Electronic Sphygmomanometer and Methods for Measuring Blood Pressure, U.S. Patent No. 4,005,701, February 1, 1977.

12. Automatic Air Deflating Regulator for Use in an Instrument for Measuring Blood Pressure, U.S. Patent No. 4,198,031, April 15, 1980.

13. Process for Coating Optical Fibers, U.S. Patent No.

4,402,993, September 6, 1983.

14. Apparatus for Coating Optical Fibers, U. S. Patent No. 4,530,750, July 23, 1985.

15. Shelf Track Lighting, U. S. Patent No. 5,034,861, July 23, 1991.

16. Disposable Microwave Package Having Absorber Bonded to Mesh, U. S. Patent 5,075,526, Dec. 24, 1991.

17. Refrigerator Water Filter, U. S. Patent 5,135,645, Aug. 4, 1992

18. Method and Apparatus for Removing Radon Decay Products from Air, U.S. Patent 5,277,703, Jan. 11, 1994

19. Hand Dryer, U.S. Patent 6,038,786, Mar. 21, 2000

20. Method and Apparatus for Providing Secure Time Stamps for Documents and Computer Files, U.S. Patent 6,209,090

21. Warning System and Method for Detection of Tornadoes, U.S. Patent 6,232,882, May 15, 2001

22. Emergency Light Device, U.S. Patent 6,336,729, January 8, 2002

23. Electronic illuminated house sign, US Patent 6,367,180, April 9, 2002

24. Electronically-Controlled Shower System, U.S. Patent 6,438,770, Aug. 27, 2002

* Provisional Patent Application for "Method and Device for Increasing the Life of Motor Brushes and

Motor Commutators"

* Provisional Patent Application for "Method and Apparatus for Bird Baths Resistant to Freezing"

* Provisional Patent Application for "Method and Apparatus for Enhancing Stress Detection Using Side Tasks"

**. Provisional Patent Application for Improved Fluorescent Light Sources

**. Patent Application for Improved Fluorescent Light Sources

**. Patent Disclosure for Image Superposition with Background Rejection or Modification

**. Patent Disclosure Document for improved source for deposition of artificial diamond and other materials, without use of vacuum systems.

**. Patent Disclosure Document for modified diamond coatings for erosion protection.

Plus many confidential patent disclosures for various clients.

# # #

Sol Aisenberg
Tel: 508/651-0140
E-mail: solaisenberg@comcast.net
Web site: http://inventing-solutions.com

Appendix A3

Curriculum vitae

Sol Aisenberg, PHD, Technology and Business Advisor In Technology Evaluation, Enhancement, Protection, And Effective Technology Transfer

E-mail: solaisenberg@comcast.net
Web site: http://inventing-solutions.com

Dr. Aisenberg specializes in the area of helping companies grow through the introduction, development, and improvement of new technology, products, and processes. He also reviews patents and inventions to help get better patent protection for clients, to make them bulletproof, and to bypass competitive patents.

He is a physicist, scientist, inventor, manager, and consultant in development and transfer of selected Technology, and Products. He has held part time positions as a Staff Member at M.I.T, and as a Lecturer at Harvard Medical School, and as a part time Visiting Research Professor in the Biotechnology Department of Boston University.

Aisenberg consults on evaluation of technology, products, and patents for clients. Practical experience as Division President, General Manager, and Principal Investigator for high technology divisions of two Fortune 500 companies. Experience includes development of lasers, medical devices, medical instrumentation, microcomputer devices, and application software.

Experienced in instrumentation development, lasers, thin films, advanced materials, plasma physics, plasma diagnostics, high vacuum systems, analog and digital circuit design, rf devices, electro-optics, RF, microwaves, and sensors. Clients include Fortune 500 corporations, small companies, patent firms, and universities. Pioneer in development, demonstration, science, technology, and applications of diamond-like carbon film (without use of hydrogen or hydrocarbon content), and in ion assisted film deposition.

Education:
PHD, Massachusetts Institute of Technology (Major: Physics, Minor: Math)
B.S., Brooklyn College, (Major: Physics, Minor: Math). Graduated Cum Laude, Honors in Physics. Phi Member of Phi Beta Kappa, Sigma XI (Science honor society), Pi Mu Epsilon (Math honor society)

Some Current and Past Professional Activities and Affiliations:
American Physical Society (APS)
Division of Astrophysics
Division of Biological Physics
Topical group in Gravitation
Division of Material Physics (APS)
Division of Plasma Physics (APS)
Institute of Electrical and Electronic Engineers (IEEE)
Instrumentation and Measurement Society of IEEE
Society of Photo-Optical Instrumentation Engineers (SPIE)
Association for Research in Vision and Optics (ARVO)

Past Member, Steering Committee, N.E. Chapter, Thin Film Div., AVS
Past Member, New York Academy of Sciences
Past member, American Inst. of Aeronautics and Astronautics
Past Member of Long Range Planning Committee of Society of
Photo-Optical Instrumentation Engineers (SPIE)
Past member, Aerospace Technology Committee, AAMI
Past member of Steering Committee, TUFTS/SPIE Engineering Update Series in Electro-Optics
Past member of AAMI Sphygmomanometer Standards Committee
Past member of AAMI Manual Sphygmomanometer Standards Subcommittee
Past member of Association for Computing Machinery
Past member of American Vacuum Society (AVS)
Past member of Plasma Science and Technology Division, AVS
Past member of Thin Film Division, AVS
Past member of Division of Electronic Materials, AVS
Past member of Materials Research Society
Past member of Computer Society of the IEEE
Past member of Association for Advancement of Medical Instrumentation (AAMI)

Served on grant review committees for NSF
Served on grant review committees for NASA
Technical Area Chairman, American Carbon Soc., 1978-79

Past Member of Licensing Executives Society
Past Member of Technology Transfer Society

Past Member of Board of Directors of the Society of Professional Consultants
Past Vice-Chairman of IEEE Consultants Network

Reviewer for:

Journal of Applied Physics; Applied Physics Letters; Journal of Vacuum Science and Technology; and Review of Scientific Instruments Listings in various editions of:

American Men of Science, Who's Who in the East, Dictionary of International Biography, Leaders in American Science, Who's Who in Finance and Industry, Who's Who in the World, Who's Who in America, and Who's Who in Technology Today.

Awards:

1964: Best Original Research Paper, "An Optical Study of Performance Limitations of Plasma Lasers", presented at the 1964 National Electronics Conference.

1970: IR-100 Award for Ion Beam Deposition Technique for Insulating Carbon Film. (Diamond-like carbon)

1971: IR-100 Award for "MICRO-PORPH" Lead Poisoning Detector.

1972: IR-100 Award for "DURA-SHIELD" Scratch Resistant Coatings.

1972: IR-100 Award for "DRUG SCREEN" for drug detection.

1972: IR-100 Award for "IMPLANTABLE GLUCOSE DETECTOR."

1973: IR-100 Award for "PULSE WATCH."

1973: IR-100 Award for "ALCOHOL BREATH SENSOR"

Experience:

1987-now   International Technology Group, Natick, Mass. Founder and President. Technology and business advisor. Develops, and consults on advanced technology for U.S. and International companies. Consults on the development of client technology. Selects, evaluates, and refines technology and products. Developer of inventions and products for licensing. Activity includes plasmas, medical devices, instrumentation, application software, and new products. Innovates technology and applications of diamond-like carbon and artificial diamond. Evaluates and develops inventions for clients. Active in technology transfer.

2004-now   Added activity as co-founder of Intellectual Property Providers (IPP) providing inventions devised by request at no cost or obligations in response to accepted requests. Inventions are then available for licensing or purchase

1999-2001   Senior advisor on intellectual property for a second major financial information company. Involved in identifying company intellectual property to be protected by additional patent applications. Review patents of competitive organizations in order to devise ways of bypassing their claims.

1999-2001 Senior advisor on intellectual property for a major financial information company. Involved in identifying company intellectual property to be protected by additional patent applications. Review patents of competitive organizations in order to devise ways of bypassing their claims.

1996-1997 Visiting Research Professor, Department of Biomedical Engineering, Boston University, Boston, Massachusetts. Involved in computer, software, electronic and optical aspects of aids for the disabled.

1990-now Invent Resources Inc. Co-founder, Director, and Vice-president. This group of experienced and productive inventors (with a combined total of over 91 patents issued, and more than 49 inventions and products licensed) works as a group to provide inventions on demand and at no cost for the inventing. These inventions are then available for licensing. Members also work with licensors to develop and improve these inventions.

1989-1990 Cell Control, Inc., Cambridge, Mass. President and Director. Develops and licenses biochemical's and delivery systems for control of human and animal cells infected with viruses and bacteria.

1987-1990 Science and Technology Resources, Inc., Framingham, Mass. President and founder

Consulting and product development

1986-1987   Appointment as a Lecturer at Harvard University Medical School

1984-now   Data Associates, Framingham, Mass. Founder, and President Developer of application software, technical reports, and time series and econometric forecast reports

1984-1987   Applied Science Group, Inc., and Applied Science Laboratories (ASL) Division, Waltham, Mass. (ASL was purchased from Gulf+Western Manufacturing Company). Executive Vice President, Chief Scientist, Principal Investigator, Co-Founder and Director. Responsible for engineering, research, development, and new products

1977-1984   Applied Science Laboratories, Waltham, Mass., a division of Gulf+Western Manufacturing Company. (Formerly Space Sciences Division of Whittaker Corporation and purchased by Gulf+Western in Jan. 1977) President, Technical Director, and Principal Investigator Responsible for company management, budgets, business plans, and technical programs as well as hiring department managers and technical staff Obtained and negotiated contracts Performed and directed research in: thin film deposition; diamond-like carbon; optical fiber hermetic coatings; optics; medical devices; bio-compatible materials; fuel cell sensors; blood pressure and heart rate measurement;

medical instrumentation; microcomputer instrumentation and software.

1964-1977   Space Sciences, Waltham, Mass., a Division of Whittaker Corporation. President, General Manager, and Principal Investigator. Progressed from Senior Scientist to Program Manager, to Physics Department Manager, to Principal Investigator, to Vice President, to General Manager, and to President of Space Sciences Division. Selected and hired marketing, technical, and financial department managers. Performed and managed research in: plasmas; thin films; medical instrumentation; optics; fuel cell sensors; and diamond-like carbon. Prepared annual business plans for Division.

1956-1964   Raytheon Research Division, Waltham, Mass. Joined as part time Associate Research Staff Member while finishing PHD thesis, and subsequently became Senior Research Scientist in the Solid State Physics Group, and then in the Theoretical Physics and Laser Group. Active as internal consultant to other Raytheon divisions. Performed research in plasmas, thin films, gas lasers, laser diagnostics, microwaves, and special materials.

1951-1956   Research Laboratory for Electronics, and Physics Department, MIT, Cambridge, Mass. Staff member and Research Assistant. Performed research in physical electronics, Langmuir probe plasma diagnostics, plasma physics, and ultra-high vacuum.

Publications:

Over 134 publications, presentations, and reports, including 17 dealing with electro-optics, 40 with instrumentation and medical devices, 91 involving thin films and plasmas, and 42 dealing with artificial diamond material, diamond-like carbon and related thin film technology. Also included are 8 awards, and many invited lectures, plus over 26 issued U.S. patents and various patents for clients.

Dr. Aisenberg has presented invited papers in the fields of diamond-like carbon, lasers, energy conversion, plasmas, thin films, medical devices, instrumentation, and superconductors.

# # #

Patents and Patent Disclosures with S. Aisenberg as Inventor or Co-inventor

1. Apparatus for Measurement of Plasma Conductivity, U.S. Patent No. 3,525,931, August 25, 1970.

2. Magnetohydrodynamic Generator, U.S. Patent No. 3,660,700, May 2, 1972.

3. Extrusion System, U.S. Patent No. 3,675,451, July 11, 1972.

4. Doppler Shift System, U.S. Patent No. 3,795,448, March 5, 1974.

5. Blood Glucose Monitoring System, U.S. Patent No. 3,837,339, January 6, 1975.

6. Apparatus for Film Deposition, U.S. Patent No. 3,904,505, September 9, 1975.

7. Perforated Wall Hollow-Cathode Ion Laser, U.S. Patent No. 3,931,589, January 6, 1975.

8. Dual Field Excitation for a Carbon Dioxide Laser, U.S. Patent No. 3,943,465, March 9, 1976.

9. Film Deposition, U.S. Patent No. 3,961,103, June 1, 1976.

10. Apparatus for Measuring Alcohol Concentration, U.S. Patent No. 3,966,579, June 29, 1976.

11. Noise Rejecting Electronic Sphygmomanometer and Methods for Measuring Blood Pressure, U.S. Patent No. 4,005,701, February 1, 1977.

12. Automatic Air Deflating Regulator for Use in an Instrument for Measuring Blood Pressure, U.S. Patent No. 4,198,031, April 15, 1980.

13. Process for Coating Optical Fibers, U.S. Patent No. 4,402,993, September 6, 1983.

14. Apparatus for Coating Optical Fibers, U. S. Patent No. 4,530,750, July 23, 1985.

15. Shelf Track Lighting, U. S. Patent No. 5,034,861, July 23, 1991.

16. Disposable Microwave Package Having Absorber Bonded to Mesh, U. S. Patent 5,075,526, Dec. 24, 1991.

17. Refrigerator Water Filter, U. S. Patent 5,135,645,

Aug. 4, 1992

18. Method and Apparatus for Removing Radon Decay Products from Air, U.S. Patent 5,277,703, Jan. 11, 1994

19. Hand Dryer, U.S. Patent 6,038,786, Mar. 21, 2000

20. Method and Apparatus for Providing Secure Time Stamps for Documents and Computer Files, U.S. Patent 6,209,090

21. Warning System and Method for Detection of Tornadoes, U.S. Patent 6,232,882, May 15, 2001

22. Emergency Light Device, U.S. Patent 6,336,729, January 8, 2002

23. Electronically-Controlled Shower System, U.S. Patent 6,438,770, Aug. 27, 2002

* Provisional Patent Application for "Method and Device for Increasing the Life of Motor Brushes and Motor Commutators"

* Provisional Patent Application for "Method and Apparatus for Bird Baths Resistant to Freezing"

* Provisional Patent Application for "Method and Apparatus for Locating and Monitoring Persons in Dangerous Situations"

* Provisional Patent Application for "Method and Apparatus for Enhancing Stress Detection Using Side Tasks"

Plus many confidential patent disclosures for various clients.

This list is not completely up to date.

Appendix B

Copy of earlier analysis of universe that was posted on my web site, "http://inventing-solutions.com" in around October 14, 2004, with slight updating of my email addresses, and before my recent simplifying insight about gravity without a power series expansion, and the red shift extrapolation.

A Simplified Model of the Universe
Clarifying dark matter, dark energy, and the big bang
Sol Aisenberg, PHD
International Technology Group
508/651-0140
solaisenberg@comcast.net
saisenberg@alum.mit.edu

Draft 9.2.4
October 12, 2004

Contents

4. Why there is a need to reexamine gravity in our solar system

5. Explaining the effect of the additional gravity contribution on the red shift

6. Describing three additional contributions to red shifts – which are independent of velocity

7. Why the Hubble assumption relating red shift to receding velocity is wrong

8. Explaining the decreasing value of the measured Hubble constant

9. Explaining tired light

10. Explaining Olbers' paradox and the dark sky

11. Explaining the distribution of stars and galaxies and quantified values of red shifts

12. Explaining why some galaxies apparently travel transversely faster than that of light

13. Explaining the different distances for connected galaxies

14. Explaining the unusually large energy output of quasars

15. Discussing black holes and information leakage

16. Explaining the growth rate of galaxies and formation of strings of galaxies

17. Clarifying the question of the age of the universe and the event horizon

ABSTRACT

Many of the mysteries in the current model of the universe are caused by the ASSUMPTION that Newton's gravitational constant can be used outside our solar system. Newton's theory of gravity is only based upon observations in our solar system and is not a universal theory of gravity as is generally believed. When we examine the equation balancing the centrifugal force against the gravitational G force in spiral galaxies, we find the basic equation $M*G = r*v*v$ where r is the distance of the rotating stars from the internal mass M, v is the rotation velocity. The excellent observations of Vera Rubin, confirmed by others, showed flat (constant) velocity curves in spiral galaxies, by the use of differential red shifts. Because of the constant velocities the result is that $M*G$ is a linear function of r, The commonly accepted belief is that the linear distance dependence is in M, because everyone KNOWS that G is a universal constant. We propose a Theory of Universal Gravity, TUG, where Newton's gravitational constant Gn can be made universal by adding a term $A*r$ to include a gravitational component linear in the distance r, which is significant at galactic spacing but hard to detect in our solar system. However, its very small effect in our solar system can be detected from sensitive, precision data from the "Pioneer Anomaly for the NASA space probes Pioneer 10 and 11, and this also explains the anomaly. Instead of spending years and money searching for invisible dark matter, we can accept the addition of an invisible linear term in the gravitational constant, which itself is invisible and is only detected by its effect on visible planets and stars.

Our theory is different from the MOND theory of Milgrom, which involves acceleration and interpolation between limits. The addition of an expansion term depending upon distance, r, - which we call an Expanded Gravitational Constant (AGC) - to Gn can predict and explain the observed flat velocity rotation curves of spiral galaxies (Rubin), and in the 1930s the unexplained motion of groups of interacting galaxies (Zwicky) - without requiring dark matter. Other mysteries are clarified when we examine the ASSUMPTION that the Doppler effect can use the Hubble constant and the red shift to measure the expanding velocity.

The red shift actually includes three other contributions that are not related to velocity, but depend upon gravity and distance. These three gravitational contributions to the red-shift will predict and explain other mysteries including the decrease in the Hubble constant for galaxies at larger distances, and Olbers' paradox (why the sky is dark). They will explain tired light by long-range gravitational drag (not scattering collisions or short range drag), and will predict the existence and uniform temperature of the CMB.

The red shift can no longer be used to support the theory of the rapidly expanding universe, the accelerating expansion, the inflationary universe, dark energy, the big bang, or the use of the Hubble constant in determining the age of the universe and the event horizon. Our simplified model of the universe explains and is supported by published observations of others.

SUMMARY

In considering many of the mysteries of the accepted model of the universe, my intuition was that a cause might be the implicit ASSUMPTION that the laws of gravity are also applicable far outside our solar system. An unexpected consequence was the realization that the Hubble law was also based upon an ASSUMPTION that the red shift could be used to measure the expansion of the universe. For the last seven decades, these assumptions have resulted in the need to invent the mysterious dark matter and dark energy to explain the observations.

There have been many who have questioned some aspects of the standard theory of the universe but without acceptance. However, my simplified model of the universe (yet another model) clarifies an unusually large number of mysteries, observations, and conclusions and may bring us closer to the real model.

We will correct two ASSUMPTIONS, that at galactic distances, the gravitation constant can be used unchanged, and that the red shift can be used to measure velocity and distances. These assumptions are responsible for many mysteries in our universe.

One of the major mysteries is the need for vast amounts of dark matter to explain the observed flat velocity rotation curves of spiral galaxies as reported by Vera Rubin and others. An even earlier suggestion for the need for dark matter was reported by Fritz Zwicky who observed the motion of galaxies in groups, and suggested that additional invisible matter was needed to satisfy the Virial theory (involving kinetic energy and potential energy).

If dark matter is believed to exist, one should explain why this dark matter is not visible from reflected light from nearby visible stars, like our moon and other planets in our solar system are illuminated by our sun. There is also the need to explain why dark matter, if it existed, did not eclipse light from nearby stars. Dark matter is presumably only based upon extra gravitational effects.

There is a simple solution to the missing dark matter and we show that they are just fudge factors and concepts introduced to explain the mysterious motion of stars in spiral galaxies and of groups of galaxies outside our solar system. (One of the biggest fudge factors in physics.)

In our simplified model of the universe is the generalization of the gravitational constant of Newton, Gn, into a power series in distance, r, and where the contributions to the gravitational force become apparent at galactic separations. Thus we suggest that there is an Additional Long Range Gravitational force, (EGC), and the gravitational constant can be represented as: $G = Gn + A*r + B*r*r$, which reduces to Newton's gravitational constant Gn at planetary separations. When asked where the coefficients A and B come from, the answer is that they came from the same place as G. Comparison with observations can show if A or B are zero.

The value of the A coefficient can be determined from data for rotational velocities of the outer portions of spiral galaxies, and is able to explain these observations without requiring dark matter. This suggests a similar explanation for the observations of Zwicky for groups of galaxies.

This additional long-range gravitational force, EGC, is different from the interesting MOND theory of M. Milgrom, which involves acceleration and interpolation between limits.

The coefficient B suggests a force independent of distance and can explain the reported tiny central attraction force experienced by the space probes Pioneer 10 and 11 in our solar system. (Anderson) The implications of this concept should be investigated but additional related data probably will be difficult and expensive to obtain.

Note that the laws of gravity by Newton and Einstein were based and validated only from observations in our solar system. It is incorrectly ASSUMED (without proof) that Newton's gravitational constant is also valid at large galactic separations.

Our theory of a generalized gravitational constant has unexpectedly led to considerations of other mysteries in the accepted model of the universe and analysis of the meaning and application of the red shift. The additional loss of photon energy (red-shift) due to moving against the additional long-range gravity was considered next.

When we examined the early work of Hubble with respect to the red shift we found that he originally described the red shift as "apparent velocity." We examined Hubble's law and the use of red shift to measure velocity and questioned its use in demonstrating the expansion of the universe. We show four contributions to the red shift and only one is related to the Doppler effect and velocity. One consequence is our prediction and explanation for the decrease of the original Hubble constant to an asymptotic value as data for further galaxies are added.

Also because the red shift contains three contributions that are not related to velocity, the use of the measured red shifts to describe the speed of the expansion will give false values for the speeds of the expansion that are larger than the actual values.

Einstein has described the effect of gravity on light photons, and the loss of photon energy (red-shift) in leaving the gravitational attraction of mass, and the bending of light by gravity. Because in our model the gravitational force extends much further than the usual inverse square component, we predict that the photon energy loss persists longer with separation and can provide an additional contribution to the red-shift continuing to very large distances – and in addition to the Doppler Effect.

Following up the long range gravitational force equation by integrating to determine the energy expended going against additional long range gravity shows that the red-shift (reduction of photon energy) in moving against this long range force gives a dependence of the red-shift as an $\ln(r)$ logarithmic function of separation. For smaller separations, this provides a linear dependence of the red shift with distance (the linear Hubble law dependence).

According the Einstein's General Relativity, one contribution to the red shift is the photon energy lost due to leaving the inverse square gravitational well of the emitting star and the associated masses. For very large masses the red shift can be significant compared to any Doppler contribution.

In searching the literature I found that Fritz Zwicky suggested the concept of "tired light" due to gravitational drag, which would explain the red shift as a linear function of distance traveled. This contribution to the red-shift becomes even more than significant because of the draining of photon energy loss due to additional long range gravitational drag on vastly more gas and dust in the interstellar path, without requiring or involving less frequent inelastic scattering collisions which would degrade the image of the source. The additional attractive gravitational force and loss of photon energy decrease inversely with distance, but the volume of interaction increases as the square of distance. Also additional dust and gas are involved in gravitational drag because of the expanded gravitational constant and range. The vastly larger number of dust and gas entities involved in the gravitational drag will result in very low directional scattering (compared to collisions) so that the image is not blurred.

The Hubble law and the red shift are assumed to measure receding velocity due to the Doppler effect. However there are three other contributions to the red shift that depend upon gravity and do not depend upon velocity.

This new understanding also explains Olbers' paradox (why the sky is dark). It is due to the loss of photon energy in traveling large distances. The number of stars increases with distance, but the light from these stars will lose energy with distance and will decrease to values below the visible range.

Thus the theory of the inflationary universe, which depends upon red shifts, should be reexamined. Also, the use of the Hubble constant, based upon red shift and velocity, in computing the age of the universe is questioned. Age determination due to nuclear processes can give the age of the oldest stars.

With the questioning of the ability of the Hubble red shift to support the concept of the rapidly expanding universe, the related mysteries such as dark energy, negative gravity, and the acceleration of the expansion all no longer are supported by observations.

The long-range $\ln(r)$ component of the red shift predicts an apparent acceleration of the supposed expansion of the universe at large values of Z, the red-shift parameter. (Perlmutter)

Reported gaps in the red shift are also explained because of gaps in distances in arrays of galaxies.

The big bang theory, which is an important part of the standard model of the universe, should be reconsidered. The concept of inflation should also be reexamined.

The observed uniformity, thermal equilibrium, and the black body temperature 3.7 deg K of the cosmic microwave background, CMB, are also predicted and explained without requiring the big bang and inflation. Our different theory, for the cosmic microwave background

(CMB) and the temperature of the CMB, is based upon the effect of gravity on photons traveling from stars at great interstellar distances and interacting gravitationally with cold interstellar dust and gas.

Thus the big bang and the inflationary universe, which are based upon the cosmic microwave background, CMB, the event horizon, the red shift and the Doppler effect, apparently lose their validity. The event horizon, which is an important part of the big bang theory, is computed from the age of the universe and the velocity of light and should be reconsidered. The concept of inflation is related to the concept of the big bang, and also should be reexamined. There are many other implications of my simplified theory of the universe. The future work of others can be more productive if they are alerted to the existence of this simplified model.

INTRODUCTION

This analysis describes my modified theory of gravity for the Universe that agrees with the gravitational theory of Newton and of Einstein - for smaller separations such as in our solar system - and is also valid and significant at galactic separations.

Presentation of my theory of the Expanded Gravitational (EGC) can simplify the model of the universe. In order to prevent this theory from being classified as speculation, items of supporting information based upon published observations by others will be included. A number of related comments will be included.

My model is different from the interesting MOND theory of Milgrom involving modified gravity. It does not involve the MOND model, which uses acceleration plus a non linear dependence and interpolation between limits.

My suggestion is that many of the beliefs about the universe are wrong and needlessly complicated and use the invocation of massive fudge factors such as dark matter and dark energy. I need only one simple assumption about gravity and components of the red shift to simplify the model of the universe and it is consistent with observations. Of course, I take responsibility for any of my errors.

The proposed model of the universe does not need to involve quantum theory or Einstein's Relativity. They are left to the many scientists who use them to explain many of the mysteries of the universe. However Einstein's theory of General Relativity involving the effect of gravity on light photons is involved and is important in explaining additional contributions to the red shift.

In our work, observations and experiments take priority over theory, particularly when there are conflicts.

For over 70 years (starting in about 1930) the scientific community has been concerned about problems and mysteries in the understanding of the universe. One problem is finding dark matter, many times larger than the visible matter represented by light from stars. Other problems are the apparent expansion of the universe and the apparent acceleration of the expansion, and the postulated dark energy and negative gravity.

According to my new theory, the problems are caused by two fundamental ASSUMPTIONS, which are commonly used by others without proof. One is that the attractive force of Newton is valid without modification at very large separations outside our solar system. The second ASSUMPTION, again made without proof, is that the red shift and the Hubble constant can be used to measure the velocity of remote stars and very far separations. This assumption about the red shift as a measure of receding velocity is serious because it has produced a belief that the universe is expanding, that the expansion is accelerating, and that there is dark energy.

Associated with the assumption of the rapidly expanding universe is the concept of the big bang, partly supported by the Cosmic Microwave Background (CMB) and the observed temperature and uniformity of the CMB.

Also part of the standard theory is the concept of the inflation phase of the beginning of the universe. The event horizon based upon the estimated life of the universe and the velocity of interaction limited to c requires the big bang and inflation to explain the uniform cosmic microwave background and the uniformity temperature. Our simplified model explains the cosmic microwave background and the uniform temperature without requiring a big bank or inflation.

To differentiate our new simplified model from pure speculation, we will provide a number of items that provide supporting arguments for our new theory. They are based upon the vast body of astronomical observations reported by others because I do not need or have access to observational equipment.

As part of my consideration of the effect of my new theory, I will identify and suggest some tasks that can be done by others who have access to primary astronomical data, or who can carry out calculations to evaluate the suitability of my theory.

Initially, starting in 1998, my analysis of the constant velocity rotation curves of spiral galaxies as reported by Vera Rubin and others has lead to my new theory that Newton's gravitational theory and the gravitational constant, G, has an additional attraction term that increases with separation r. It provides a simple extension of the gravitational force of Newton and Einstein that is only significant for large separations. Unexpectedly it leads to understanding of many other mysteries of the universe.

My new theory is that the gravitational constant, G, can be generalized and expanded into a simple power series in terms of distance, r, and in the simple form $G = Gn + A*r$ where Gn is Newton's gravitational constant and where A can be proven to be non zero when evaluated using observations of spiral galaxies. When asked where the term $A*r$ comes from, an answer can be that it comes from the same place as Newton's gravitational constant.

Thus according to my new theory, the inverse square attractive force between masses is augmented at very large separations by another force that decreases much slower as a function of separation and involves an additional term in describing the gravitational constant. The new representation of the long range gravitational constant Ga is approximated as part of a power series: Ga = Gn + A*r where Gn is Newton's gravitational constant, r is the separation, and the constant A is to be determined by use of published observations - or else shown to be zero. Note that this reduces to Newton's and Einstein's description of gravity in our solar system where separations are smaller than galactic separations.

With very precise measurements possible using space probes, a very tiny addition to gravity apparently can be detected and measured, and this alone shows that the theory of gravity should be reexamined. Note that there may be a second addition in the form of B*r*r which may explain the motion of the space probes Pioneer 10 and 11 in our solar system. The implications of this term may be studied later.

Others have suggested a modification of Newtonian gravity. One significant suggestion was MOND (Modified Newtonian Dynamics) hypothesized in 1983 by Moti Milgrom. Briefly, in the MOND version the modification of the effect of gravity occurs at very small accelerations, involves nonlinear acceleration terms, and uses interpolation functions. It is different from our simple generalization of the gravitational constant that adds a few terms of a power series that depends upon separation and does not involve acceleration.

Implications of our simplified theory influence and explain the supposed expansion of the universe, the acceleration of the expansion of the universe, negative gravity, dark energy, the big bang, and the deduced transverse velocity of very remote galaxies at velocities greater than the velocity of light. A significant contribution is the prediction of the decreasing Hubble constant, and the understanding of the true meaning of the key red shift observations

An important part of the value and validity of a new theory is the ability to agree with existing observations, but even more important to make many predictions that can be confirmed by future observations.

We will provide descriptions of observations that can be explained and predicted by our theory of galactic gravity. Additional items will be added from time to time.

We provide 21 items at present.

SUPPORT FOR OUR THEORY, FROM OBSERVATIONS OF OTHERS

ITEM #1
EXPLAINING THE APPARENT DARK MATTER IN SPIRAL GALAXIES

While observing a Public Broadcast TV program describing the flat rotation velocity curves of spiral galaxies I found it hard the believe the postulated existence of massive amounts of dark matter. Instinctively I felt that if, in the region showing constant rotational velocity curves, the relationship between gravitational attraction and distance was an inverse function of distance rather than the usual inverse square relationship (established in our solar system), the observed flat velocity rotation curves could be explained without needing the massive fudge factor of dark matter.

The outward force on a mass, m, rotating at a tangential velocity, v, at a constant radius r around a central mass M is balanced by the gravitational attraction and is described by the equation:

$$m*v*v/r = m*M*G/(r*r)$$

this reduces to

$$v*v*r = M*G$$

We can assume that M is a function of r, or G is a function of r, or both are, giving

$$v*v*r = M(r)*G(r)$$

Thus for the outer region of spiral galaxies where the red shift data show a flat velocity rotation curve as a function of radius, we are led to the conclusion that in the outer region of the spiral galaxies the product of G and M is a linear function of radius. If one ASSUMES that the gravitational constant G is independent of distance, r, then one is led to the usual conclusion that the mass M must be a linear function of r in that region, even if such mass is dark and not visible. The amount of dark matter required to satisfy the observations is said to be much larger than the visible matter.

By taking the alternate explanation where the gravitational constant G is expanded as a function of r, and has a component that is linear in r, we can explain these observations without dark matter. When this component is significant at separations outside our solar system (at separations comparable to the size of spiral galaxies, about 3 kilo parsec) we can explain the observations without needing dark matter. This alternative is desirable according to William of Ockham's Razor, which prefers the simpler alternative.

There is a need to explain why dark matter, if it existed, is not visible from reflected light from nearby visible stars, like our dark moon is illuminated by our sun. There is also the need to explain why dark matter, if it existed, did not eclipse light from nearby stars like eclipses caused by our moon and the inner planets. Dark matter is presumably only based upon extra gravitational effects. Our theory of a gravitational constant that has a component that increases slowly with distance eliminates the need for massive dark matter and the associated mysteries. The work on searching for dark matter such as WIMPS can be redirected.

Published data for flat velocity rotation curves of spiral galaxies helped determine an initial estimate for the value for A. This constant, A, describing the additional long-range gravitational force, was evaluated by using the published observations for the constant velocity outer rotation curves of spiral galaxies. The constant, A, was evaluated to be none zero because it described the observed motion without needing dark matter.

At the transition radius for spiral galaxies, Rs, where the rotation velocity curves become constant, the Newtonian force and the additional long range force become approximately equal. An estimate of Rs was obtained from the intersection of an extrapolation of the rising portion of the rotation curve with the extrapolation of the flat portion of the rotation curve. Preliminary analysis of data from spiral galaxies NGC2403 and NGC3198 using a spiral Galaxy transition radius Rs of 2.7 Kparsec plus the known value of Newton's gravitational constant Gn (Gn = 6.672 x 10^(-8) cm*cm*cm/gr/sec*sec) gives a preliminary value for A = Gn/Rs = 2.16 10^(-26) cm*cm/gr/sec*sec). This should be refined by curve fitting using data for rotation velocity as a function of radius, r, for a number of spiral galaxies.

Thus this theory, based upon published observations, shows that dark matter is not needed to explain flat rotation curves of spiral galaxies.

Unexpectedly, this has permitted me to explain and predict many other observations and mysteries in the universe - and could lead to questioning many well-accepted theories of the universe. These additional insights will be described in later pages, in web sites, and possibly in some professional publications.

Item #2
EXPLAINING EARLIER SUGGESTION OF DARK MATTER BASED UPON MOTION OF GALAXIES IN GROUPS

The unusual motion of remote groups of galaxies was earlier described by Fritz Zwicky (about 1930), and introduced the concept of dark matter (missing matter). Zwicky used observations of the velocities of galaxies in a group of galaxies together with the observed separations of the galaxies in the group. The Virial theory showed a discrepancy between observations and theory. One term in the Virial equation involves kinetic energy, and the other term involves potential energy (using the gravitational constant). Zwicky suggested the existence of large quantities of dark matter to potentially resolve the discrepancy.

Thus the concept of dark matter was introduced as an explanation for unusual galaxy motion, and was prior to the concept of dark matter proposed for spiral galaxies. It is suggested here that the Virial theory involving kinetic energy and potential energy (involving a gravitational constant) should be reexamined for cases of galactic separations including an expanded gravitational force. The Virial theory should be valid unchanged when used to determine the mass of planets rotating around a sun because of the much smaller separations.

Thus, observation of unusual motion of groups of spiral galaxies can support our model of the long-range gravitational constant. Because of our simple model involving an additional long-range gravitational force, the unusual motion of groups of galaxies was explained and predicted without the need to invoke massive amounts of dark matter. It provides a beautiful alternative to dark matter. Again according to William of Ockham's razor, the simplest explanation is preferred when it is consistent with past observations and with future predictions.

Item #3

EXPLAINING WHY THE GRAVITATIONAL THEORY OF NEWTON AND EINSTEIN ARE STILL VALID IN OUR SOLAR SYSTEM

Newton's theory of gravity is based upon observations in our solar system and therefore is still valid in the solar system. The proofs of Einstein's theory of General Relativity are also based upon observations in our solar system.

Four proofs of General Relativity are:

Rapid precession of Mercury's orbit
Bending of light passing near sun and influenced by gravity
Gravitational red shift in strong gravitational field
Time dilation in gravitational fields – depends upon distance from center of earth.

Because the separations in our solar system are very small compared to galactic separations, the additional component of the gravitational constant is small and the description of the effect of gravity can be described by the usual inverse square dependence in our solar system.

ITEM #4
WHY THERE IS A NEED TO REEXAMINE GRAVITY IN OUR SOLAR SYSTEM

This is just suggestive and needs additional analysis. Observation involving very high precision measurements within our solar system appears to support our theory of an additional gravity contribution. Observations of Pioneer 10 and 11 probes indicated that they were slowing down faster than predicted by Einstein's general theory of relativity. "Some extra tiny force - equivalent to a ten-billionth of the gravity at Earth's surface - must be acting on the probes, slowing their outward motion." (Anderson)

Analysis by John D. Anderson and his team at JPL ruled out a number of possible explanations of this extra force. Our theory predicts a very tiny force within solar system separations, and it is too small to significantly influence the motion of planets but can slightly influence space vehicles.

The precision measurements reported for the motion of NASA probes Pioneer 10 an 11 indicated a very tiny but verified attractive force towards the sun in addition to the expected gravitational force of the sun. This indicates that there apparently is an additional term in the gravitational attraction that and becomes large enough to detect with precision measurements at solar system separations.

THE IMPLICATIONS OF THE ADDITION THE GRAVITATIONAL CONSTANT HAS LEAD TO THE FOLLOWING UNEXPECTED INSIGHTS ABOUT THE RED SHIFTS.

ITEM #5
EXPLAINING THE EFFECT OF THE ADDITIONAL GRAVITY CONTRIBUTION ON THE RED SHIFT

As an unexpected result of the enhanced gravitational theory we found that due to the additional long range attractive gravitational force there was a change of potential energy that becomes significant for light traveling large distances - and this contributes to the red shift of the light. Integrating the force over distance gives the energy change, and it shows that there is a long-range red shift related to photon travel distance in addition to the red shift and blueshift due to the Doppler effect. This is confirmed by observations showing a linear plot of red shift as a function of distance to remote stars, and where the distance is determined from observations of Supernovas Type 1a.

However, when the equation involving the galactic gravity term is integrated to describe the energy lost in traveling galactic distances the result is a term that is a logarithmic dependence on distance. For the beginning part of the logarithmic function, the energy loss (and red shift) is approximately a linear function of distance.

But for distances corresponding to light from very far light sources (stars) the distance based upon measurements of light intensity received on Earth will indicate a distance larger than that deduced from the red shift. There is an upward curvature of light deduced distance vs. red shift-deduced distance. This has been (incorrectly) used to claim that the acceleration of the universe is accelerating (Perlmutter). It also suggests dark energy and negative gravity.

I predict that when the separations (distances) for Supernovas Type 1a light sources are plotted on the y axis against the logarithm of the Z values (which correspond to red shifts), on the x axis, the upward curving portion of the plot will be reduced for the cases of very large separations. This can eliminate the need to propose an accelerating expansion. In any event the use of red shifts to measure distances should be reexamined, and can not be used to support the expansion of the universe or the acceleration of the expansion.

ITEM #6
DESCRIBING THREE ADDITIONAL CONTRIBU-
TIONS TO RED SHIFTS – WHICH ARE INDEPEN-
DENT OF VELOCITY

Observations of red shift and deduced star distances of supernovae Type Ia have demonstrated a linear relationship where the red shift increases with travel distance of the photons. Fritz Zwicky had earlier introduced the concept of "tired Light" where photon energy is lost in traveling, and due to gravitational drag. When others assumed a model where the photon energy was lost in collisions with space dust, objections were raised because this model implied diffuse scattering that would lose image quality. Also in view of the gravitational range of the classical inverse square effect of gravity, the probable density of gas and dust in space was too small to contribute much gravitational drag and red shift. However, in view of the extended range of the inverse r component of the effect of additional gravity drag, the gas and dust in interstellar space can contribute significantly to the red shift.

In my theory (particularly where the effect of gravity persists for larger separations) the mechanism of interacting gravitationally with interstellar dust extracts energy from the photon, resulting in a red shift. The dust particles are moved slightly by the gravitational force of the mass of the passing photon and this is similar to the transfer of energy from the moon to produce tidal movement in the oceans of Earth resulting in a loss of kinetic energy of the moon. Observations have shown that gravitational fields will influence the path of photons. An example is the gravitational lens effect.

When many tiny gravitational interactions are involved with interstellar dust, the average effect retains the direction of the photon, thus preserving image quality. Fritz Zwicky was indeed right about his observation of "tired light" With this insight about the cause of red shift the ASSUMPTION that the red shift was caused by the Doppler effect has led to many false directions about the model of the universe.

This suggests a TASK involving recalculation of the gravitational drag and the enhanced red shift.

There are four contributions to the observed red shifts.

First is the Classical Doppler shift due to velocity.

Second is the photon energy loss in traveling galactic distances and losing energy due to the long-range component of the gravitational constant. Note that this is a logarithmic dependence upon distance and does not involve velocity.

Third there is the photon energy loss due to gravitational interaction with dust and gases in the photon travel. This transfers motion and energy to the dust and can become significant because the interaction range is increased significantly because of the long-range gravitational component. For this component the red shift is a linear function of distance, and does not involve velocity.

Fourth is the component due to the photon energy lost in leaving massive masses due to the effect of gravity on the photon. This is related to Einstein's theory of general relativity and the effect of gravity on light. The effect of gravity on the red shift was considered by Robert Trumpler who believed that the gravitational effect could explain observed excess red shifts of some stars and could be used to confirm General Relativity. In around 1935, he calculated that the strength of gravity on the surfaces of stars was too small to give a gravitational red shift as large as observed. (Arp, 1998, p. 98) I suggest that if Trumpler had included the additional long-range gravitational component in addition to the usual inverse square gravitational term, he might have obtained calculated red shifts more in agreement with observations.

Large masses are present in black holes in galaxies. This explains observations of galaxy pairs that apparently are at similar separations (because of observed streams of stars between the galaxies) and the different red shifts that are supposed to show that they are at significantly different distances (Halton Arp). Red shift due to massive and/or small radius black holes in one galaxy can predict and explain this mystery of connected galaxies but with different red shifts. Again, this component of the gravitational red shift is not related to velocity or distance.

There is no reason to expect the receding velocity to increase with separation/distance. Thus the red shift support for a rapidly expanding universe is not valid.

An important implication of the new understanding of the red shift is that it questions the use of observed red shifts to show that the universe is expanding, or the acceleration of the expansion. It also questions the dark energy, negative gravity. Also questioned is the theory of the Big Bang and the inflationary universe - which use the supposed expansion as one of the supports.

By combining the ln(r) dependence of the red shift energy loss and the linear dependence on r due to gravitational drag (tired light), we can predict a red shift dependence of D1*r + D2*ln(r).

I have been unable to obtain access to tabulations of distance determined from light received and the associated measured red shift so that I could evaluate the coefficients D1 and D2 and determine the ability of the equation to fit the observations. The form of the function appears to fit a published graph of the data. At some point I will expand the published graph of the data (Perlmutter), and read off the coordinates for use in curve fitting.

This could be an interesting TASK for others, particularly a graduate student.

REEXAMINING THE HUBBLE CONSTANT

ITEM #7
WHY THE HUBBLE ASSUMPTION RELATING RED SHIFT TO RECEDING VELOCITY IS WRONG

When examining the history of the use of the red shift for very remote stars as a way of measuring receding velocity, we found that there was another unproven ASSUMPTION and was made by Hubble and others. We learned that the original papers (Hubble and Humason, 1931) had a footnote that indicated that it is not certain that the large red shifts should be interpreted as a Doppler effect but for convenience can be interpreted in terms of velocity and referred to as "apparent velocities." This assumption was later incorrectly converted into evidence of actual velocity and led to serious beliefs about the rapidly expanding universe and subsequently the acceleration of the expansion. It apparently even misled Einstein.

The three velocity independent contributions to the red shift question the use of the red shift to measure the velocity according to only the Doppler effect.

Also because the red shift contains three contributions that are not related to velocity, the use of the measured red shifts to describe the speed of the expansion will give values of the speeds of the expansion that are larger than the actual values.

Having opinions and wishes about a static universe or an open or closed universe can lead to assumptions and fudge factors that are not validated by observations. Observations, when interpreted in the wrong way (like the red shift) can be misleading.

ITEM #8
EXPLAINING THE DECREASING VALUE OF THE MEASURED HUBBLE CONSTANT

Another observation and problem is the large decrease in the Hubble constant as measurements are made to include the more remote stars.

Our hypotheses and the resulting equations (in Appendix A) predicts that the Hubble constant relating red shift to distance will be larger for the closer stars and will decrease to an asymptotic value for the more remote stars that can be measured with more modern techniques. Published observations show that the early Hubble constant is about a factor of five to ten greater than modern values. Re-measurement of the Hubble constant for the nearer stars should be done using modern equipment to confirm the trend downward for greater distances and to see if the data fits the equation we derived for the dependence of apparent Hubble constant on measurement distance (which increased with time and the ability to measure out to further distances). See Appendix A, equation 8.

A serious error in the current theory of the universe is the assumption that red shifts of remote stars are only due to velocity rather than including red shifts caused by travel distances. This assumption suggested the current theory of the rapidly expanding universe, plus the acceleration of the expansion. It is also related to the model of the Big Bang, the concepts of the cosmological constant and dark energy to power the acceleration of the expansion. Before we speculate too much on the origin of the universe, and the end of the universe, perhaps we should first understand the present observations of the universe.

OTHER OBSERVATIONS INFLUENCED BY THE THEORY OF ADDITIONAL LONG RANGE GRAVITY

ITEM #9
EXPLAINING TIRED LIGHT

This provides additional support by explaining the "TIRED LIGHT" concept of Fritz Zwicky, which earlier was presented without acceptable physical reasons for the loss of photon energy with travel. Zwicky was correct in proposing "TIRED LIGHT" and energy loss in connection with the red shift, and there is now a physical explanation and prediction due to photon energy loss in moving against a long range component of the gravitational constant, plus the increased gravitational drag. Zwicky should have been taken more seriously.

There is a contribution to the loss of photon energy traveling large interstellar distances. Photons are influenced by gravitational attraction to mass as has been shown by Einstein and by deflection of light by our sun, and by the observation of gravitational lenses.

When photons travel, they expert forces on dust and gas within their gravitational influence and   transfer motion end energy to the dust and gas. This energy loss is proportional to distance. Loss of energy by inelastic collisions and scattering, which would blur the image, is less likely and less important.

The increased range of gravitational attraction associated with the additional gravitational component will increase the interacting volume and the gravitational drag on the dust and gas in that volume. There is no need for the less frequent dust collisions that would also blur the images.

Thus the red shift and blueshift have a component due to gravity and the travel separation/distance in addition to the Doppler shift due to the velocity component. For large distances, the travel component becomes significant.

ITEM #10
EXPLAINING OLBERS' PARADOX AND THE DARK SKY

The meaning of the dark sky (Olbers' paradox) and the Cosmic Microwave Background (CMB) may also be reexamined from the point of view of the new explanation for "tired light." Our theory predicts that when light from very remote stars reach us, their energy loss due to the large travel distance has decreased the energy of the electromagnetic photons below that of the visible range, and where some photons are in the microwave range (CMB).

The slower photons essentially come uniformly from all directions with slight irregularities due to irregularities in the spatial distribution of stars. Thus light from some of the stars in the sky from large distances will have their wave lengths shifted by the distance traveled and down to ranges below the visible range and can only be detected by instruments that can detect infra red, radio signals, and microwaves.

When the photons from the source shift out of the visible range they are replaced by the higher energy UV photons that will red shift down to the visible range. However, the Planck black body radiation curve shows that the number of photons with high energy decrease rapidly with photon energy. At some point there are not enough high-energy photons to replace the original visible photons and the sky becomes black from the point of view of the very remote stars with higher red shift due to large distances.

ITEM #11
EXPLAINING THE DISTRIBUTION OF STARS AND GALAXIES AND QUANTIFIED VALUES OF RED SHIFTS

Another prediction of my model is that because there are voids and discontinuities in the distribution of stars and galaxies, there should also be discontinues in the red shifts due to discontinuities in distance traveled. This is because there are discontinuities in the distances traveled and the associated loss of photon energy, according to my model.

As part of this model I predict that the discontinuities and quantified values of red shifts (and possible periodicity) will be seen only when the red shift data are collected from a narrow angle view of the universe.

I also predict when red shift data are examined for periodicity for wider-angle views of the universe the periodic behavior will disappear and will be rejected because the periodicity characteristics will be different in different directions and will fill in the gaps in the periodic aspects.

An example can be seen if one takes slices through a sponge which will have a random array of pores and gaps.

When the images of the slices are overlaid, the gaps in the array will be filled in for the composite analysis.

Observation and red shift with gaps and periodicity in the gaps have been reported by Halton Arp, and is described in: Seeing Red: Red shifts, Cosmology and Academic Science, by Halton Arp, (Apeirion, Montreal, Canada) 1998

Indeed, I have read that the publication of the report of gaps in red shifts was rejected when data from larger studies showed no gaps - and this is in agreement with my prediction that when red shift scans are made for additional directions, the gaps will be overlaid and will not be observable.

It has been reported that the gaps are in the ratio 1, 2, 3. There might be a connection between these red shift gaps and the ratio of atomic number 1, 2, 3, for atoms H, He, Li in the interstellar space. This is just my speculation.

ITEM #12
EXPLAINING WHY SOME GALAXIES APPARENTLY TRAVEL TRANSVERSELY FASTER THAN THAT OF LIGHT

It has been reported (Halton Arp) that when the observed angular transverse velocity (Proper velocity) of some (presumably) remote galaxies is used in conjunction with the distances determined from observed red shifts of these galaxies, the computed transverse velocity is larger than the velocity of light. Obviously there is an error in some of the assumptions, if the observations are to be believed.

My contribution to the understanding of the additional red shift components due to gravity predicts that if a galaxy has a massive mass then the additional red shift due to the long-range gravity component will compute to give an incorrectly large distance. The observed transverse motion of the galaxies in question can be verified by repeated observations. However the separation/distance determined from the observed red shift probably is too large because of a red shift contribution due to the effect of gravity on escaping photons according to Einstein's work on the interaction of gravity and photons.

Thus the galaxy and their stars are not as far away as assumed from the measured red shift and the galaxies are not moving faster than the velocity of light, which is not reasonable

ITEM #13
EXPLAINING THE DIFFERENT DISTANCES FOR CONNECTED GALAXIES

It has been reported (Halton Arp) that some pairs of galaxies are apparently connected by observed streams of stars moving from one galaxy to the adjacent galaxy. Surprisingly, the red shifts observed for these galaxy pairs indicate a very large difference in distance from the observer, thus falsely indicating that they can not be adjacent in spite of the observed stream of stars.

My explanation is related to the various contributions to the observed red shift. The loss of energy for a photon leaving from a massive body will be added to the red shift due to distance. If the galaxy showing a larger apparent distance contains an unusually large black hole the red shift will be larger suggesting a big difference from the apparent distance of the other galaxy in the pair. Thus the mystery is explained by an understanding of the components of the red shift and the contribution of gravity to red shifts.

ITEM #14
EXPLAINING THE UNUSUALLY LARGE ENERGY OUTPUT OF QUASARS

Some quasars are supposed to be the most distant objects in the universe. The host galaxies of the quasars appear very faint, but because of the very large separation of the quasar (as determined by the red shifts) the calculated energy output of the quasar is computed to be as much as billions of stars. However, the energy output of the quasar is computed using the inverse square correction and involves the separation/distance based upon the observed red shift and using the Hubble constant.

We have shown that the measured red shifts contain contributions due to Einstein's gravitation escape energy loss, and the tired light effect due to gravitational drag, and the additional energy loss related to the expanded gravity component.

If the quasars contain very large masses (like black holes) the Einstein gravitational contribution will make the observed red shift and the separation/distance much larger than actual. The result, when the received light from the quasar is used with the erroneous separation/distance to compute the quasar energy output, the apparent energy output can be enormously wrong. The quasar is not as far as computed from the red shift.

ITEM #15
DISCUSSING BLACK HOLES AND INFORMATION LEAKAGE

This is intended to show how an understanding of gravity can explain many of the surprising aspects of the universe.

The recent article (July, 2004) in Economist.com describing the reconsideration by Dr. Hawking of his 1970 theory of the black hole will help refine our scientific knowledge.

Hawking now feels that information can escape the black hole event horizon and that the event horizon is an "apparent horizon"

Without calling on quantum mechanics, I suggest that basic physics and theory of the Schwartzchild radius (which defines the size of the black hole) can also predict and explain information leakage and in a less complex manner. I describe how some light (information) can escape the black hole if the light originates from stars above the bottom of the black hole potential well.

The escape energy required for a mass m to escape from a potential well with a mass M, and a radius r, is given by G*M*m/r where G is the gravitational constant of Newton. The energy of a light photon with mass m and velocity c is m*c*c/2. In order to escape the gravitational potential well the photon energy must be greater than the escape energy. Thus the Schwartzchild radius Rs for a light photon with velocity c is Rs = 2*G*M/(c*c).

When the Schwartzchild radius is near galaxy size the value of the gravitational constant must include the additional long-range gravitational contribution.

It can take much time for a trapped star to travel or fall the distance of the Schwartzchild radius to reach the bottom of the black hole, and some of the photons emitted from the falling star can escape the black hole because it does not have to travel all the Schwartzchild distance.

A black hole usually is within a galaxy and sucks in nearby stars. Not all the stars in the black hole are yet at the bottom of the potential well, and there will always be some stars above the bottom of the well. Stars will emit light and light photons escaping from a potential well will have their energy reduced while the velocity c will remain unchanged. When a photon loses energy the wavelength shifts to the red and the shift depends upon the distance to the "apparent horizon." A photon from a star at the bottom of the well can lose enough energy to change to almost infinite wavelength (not detectable) in reaching the Schwartzchild radius and thus can not be seen to escape. Light from stars not at the bottom of the gravitational well can escape.

Thus photons from stars not at the bottom of the black hole gravitational well will not need as much energy to escape and can leak out of the "apparent horizon." Their wavelength spectrum (uv, visible, infrared, microwave, rf, or longer wavelengths) will depend upon the location of the emitting star descending to the bottom of the black hole potential well.

Because stars in the black hole convert some of their mass to light, there is a mechanism for mass to leak out through the apparent horizon - along with information in the form of light (electromagnetic energy).

The additional long-range gravitational constant can be a factor in the attraction of mass to the black hole and the galaxy containing the black hole.

There is a possibility that when enough mass accumulates in the mass at the center of the black hole, the black hole may explode in a manner similar to those galaxies such as super nova that explode when enough additional matter is fed into it by a nearby galaxy. Long periods of time may elapse while the stars and galaxies in the black hole funnel slowly spiral down to add to the mass at the center of the black hole. Nuclear processes still occur in the stars being trapped.

When the black hole explodes, there is a release of the mass - and the information (in mangled form).

ITEM #16
EXPLAINING THE GROWTH RATE OF GALAXIES AND FORMATION OF STRINGS OF GALAXIES

When the usual gravitational constant and the inverse square distance dependence is used to compute the time for formation of stars and galaxies, there can be differences in the computation compared with other determinations of the formation time.

My theory of long-range gravity when used in the mathematical simulation of the growth of stars and galaxies should give more reliable results.

Also it has been observed that galaxies form with structure such as voids, walls, and strings of galaxies. The model of long-range gravitational constant can predict the formation of galaxies in string formation. If one has two galaxies separated by a distance comparable to the range of the inverse distance component of the gravitational force the longer range force components will overlap and in the direction defined by the line connecting the two galaxies. The total gravitational attraction will be larger in the direction of that line than in the direction perpendicular to the connecting line. Thus there will be a greater tendency for attracting other galaxies to line up in the direction of the connecting line. There will be a progressive addition of galaxies along that line. Thus a string of galaxies can form drawing on the availability of galaxies within the range of the additional gravitational force.

A similar situation can work in the formation of walls of galaxies. Of course, voids will form where the galaxies are drawn away by the strings and walls.

THE BIG BANG REEXAMINED – AND THE TEMPERATURE OF THE CMB

ITEM #17
CLARIFYING THE QUESTION OF THE AGE OF THE UNIVERSE AND THE EVENT HORIZON

The age of the universe is estimated to be about 14 billion years (R. P. Kirshner) and this is based upon the currently accepted value of the Hubble constant. If the Hubble constant is a measure of the velocity of expansion as a function of distance, and if the velocity of a red shift source was considered to be constant during its motion, then it is possible to determine the time (age of the universe) that elapsed during its motion by computing the reciprocal of the Hubble constant. However this assumes that the red shift used in determining the Hubble constant is only due to the Doppler shift, and does not have a contribution due to tired light (gravitational drag on the traveling photons).

Also, there may be a logical problem in using the inverse of the Hubble constant to determine the age of the universe. According to the Hubble concept the velocity of the receding stars increases with separation/distance, and assuming a constant velocity over the distance traveled is not logical. I believe that the equations of motion for changing velocity should be used to determine the time elapsed in moving with a changing velocity to a specific distance, rather than just taking a reciprocal of the Hubble constant – but only if the red shift depends only upon receding velocity.

The age of the universe can be determined independently by analysis of spectrographic data of the various nuclear isotopes produced by nuclear process in stars. The age of some stars determined in this way is sometimes larger than the age determined from the Hubble constant. The reexamination of the meaning of the Hubble constant can resolve the discrepancy.

The age of the universe is used to specify the range of the event horizon. This is the maximum distance that can be traversed by light or energy traveling for the age of the universe. This concept is used in the limiting thermal equilibrium involved in the theory of the big bang.

ITEM #18
PROVIDING AN ALTERNATE EXPLANATION OF THE COSMIC MICROWAVE BACKGROUND AND ITS TEMPERATURE

Observations of the presence of low level microwave radiation coming from all directions of the universe were important in showing the uniformity of the sources. The data were fit to a black body function with an apparent temperature of 2.73 deg K which is close to zero. The temperatures from different directions differ from the average by only one part in 100,000 (A. H. Guth). The concept of the big bang is accepted as an explanation for the temperature equilibrium between regions of the universe that could not interact in view of the event horizon, which is based upon the life of the universe, and the limit of the velocity of light. There is not enough time for widely separated portions of the universe outside the event horizon to interact and exchange energy. Thus we have the theory of the big bang.

The distance that the original photons that travel and lose enough energy to drop into the microwave wavelength can be estimated from the approximate wavelength of source visible light at 0.5 microns and the detected microwave wavelength of about 10 cm. The Z value is essentially the ratio of these wavelengths and this gives $Z = 2*10\,^5$ (Z= 20,000) which is a very large value indicating a very large distance of travel, enabling thermal equilibrium with gas and dust in the vast interstellar voids between stars.

In my simplified model of the universe, photons traveling a large distance (within the event horizon) will interact by gravity and extended gravity with dust and gas in its path (without needing collisions or scattering) and over the course of billions of interactions and will come into energy equilibrium with the dust and gas in the universe.

Because the dust and gas in the universe (outside of matter in stars and galaxies) are at a low temperature such as 2.73 deg K, the photon energy will not be less than that temperature, nor more than that temperature, for photons originating and interacting over large distances. Thus the CMB will be in black body thermal equilibrium with dust and gas temperature and will be uniform in all directions. Of course this does not apply to photons that arrive in the visible range because they have not traveled far enough for the photon energy to fall to the microwave energy level.

Thus we have an alternate explanation of the CMB without involving the big bang, or the concept of inflation or of the rapidly expanding universe, or the limitations of the event horizon.

ITEM #19

DEFLATING THE BIG BANG

The realization that the red shift is not solely dependent upon the Doppler effect removes support for the concept of receding stars and the rapidly expanding universe. The big bang theory is based upon the red shift and the concept of an apparently rapidly expanding universe that can be extrapolated back to a small source. It is also based upon the meaning of the CMB and the uniform black body equilibrium and the theory that the thermal equilibrium occurred when the universe was small and that the photon energy cooled down during expansion. Another support advanced for the big bang is the ability to predict the values of the lighter elements, hydrogen, helium, and lithium. We have no comment on the nuclear processes but they alone can not validate the theory of the big bang.

An argument for the big bang is the uniformity of the CMB in spite of the inability for regions of space to interact and equilibrate for distances greater than the event horizon.

Thus the theory of the big bang involves a starting universe smaller than the event horizon so that the components can exchange energy and come into thermal equilibrium. Then according to the theory, the initial universe inflates in a very short time (a very small fraction of a second) to be bigger than the event horizon, while retaining the thermal equilibrium. There is a conflict in the theory – it should be shown that the velocity of the expansion is not greater than the velocity of light in view of the expanded size of the universe and the fraction of a second available for expansion. Or are the laws of physics and the limit on the velocity of light suspended during the big bang?

There is another question. Was all the great mass of the universe present in the first fraction of a second and in a very tiny volume such as in the form of protons and electrons, or was additional mass created during the expansion? Where is the observational evidence?

If the mass of the universe was there at the start, how did the expanding mass escape the black hole gravity associated with this great mass?

According to our model of long-range gravitational attraction to cold dust and gas in the universe, the photons will come into energy equilibrium with the cold dust and gas when traveling long interstellar distances. Traveling long distances reduces the photon energy down to the microwave range thus providing a CMB. The temperature of the photons will come into equilibrium of the gas and dust in the vast spaces encountered in the travel thus providing low temperature thermal equilibrium. This eliminates the need for the big bang to explain the temperature equilibrium of the microwave photons.

ITEM #20
PREDICTION OF THE OBSERVED VOIDS IN THE MAPS OF THE UNIVERSE

Maps of the mass distribution in the universe at different cosmic ages and distances have been made using COBE, Boomerang, and WMAP surveys.

They have been taken at wavelengths ranging from 7.24 cm (Penzias and Wilson) down to 0.33 cm corresponding to the microwave range.

If we assume that the longer wavelength photons correspond to further distances from the observer then for remote voids (where there is less mass compared to the immediate surroundings) then I predict that the typical size of the voids when measured and computed will show a dependence on the measuring wavelength and distance to the voids.

The typical size can be the average area of the individual voids, or else the RMS value, or the mode as measured in units of degrees. Because of the geometry involved, the typical size should be a linear function of the distance in the maps at different wavelengths. The ratio of the 7.24 cm and the 0.33 cm is about a factor of 22 and has a potential for testing over a significant range.

I predict a linear dependence of the average void area as a function of map distance.

Note that the voids corresponding to the visible range (about 0.5 micron) can be considered as a limiting value when plotted on a graph of the average void area as a function of wavelength (which can be considered as a measure of the map distance).

Item #21
MODIFYING THE THEORY OF GRAVITY WAVES

In view of the modified value of the gravitational constant at separations outside our solar system the theory of gravity waves should take this into question if gravity waves from other galaxies are to be detected and studied.

Even for gravity waves to be detected originating from our galaxy "Milky Way" the additional component of distance dependent gravitational constant is still greater than about 10% of Newton's gravitational constant for separations greater than 10% of the radius of the galaxy.

Will the wave equation be seriously modified by the additional long-range gravity component?

The calculation of the expected gravitational signal strength should be refined to take this into effect to see if the sensitivity of the design is sufficient to detect these gravity waves.

CONCLUSION

If considered seriously, the new theory (without the need for assumptions, dark matter, dark energy, or other fudge factors) could result in a new, correct, and simplified view of the universe and could help future work of those in the field. The theory may only be accepted by a new generation of physicists. I have read that progress is "made funeral by funeral" as resistance to new theories decreases. One of my responsibilities is to make my theory available to future generations for their judgment. It sometimes takes an average of about 20 years (a generation) for new ideas to be accepted.

It may make future research work more productive, and can lead to additional sponsored work for others in the field.

Remember, according to William of Ockham's razor, the simplest explanation is preferred when it is consistent with past observations and with future predictions.

A draft of a prior version of my model is provided as a preprint including details of the new theory and many more implications are presented at:

http://inventing-solutions.com/new-universe.htm.

The second version of this model including details of the theory document is provided, at:

http://inventing-solutions.com/new-universe-theory. htm

The web site for this latest expanded present document includes extensions and revisions of the previous two web postings and is located at:

http://inventing-solutions.com/simplified-universe. htm

APPENDIX: A - EQUATIONS

This provides the evolution of the equations used for the analysis based upon the theory that includes a long-range term, A*r, to Newton's gravitational constant Gn.

FOR ROTATION CURVES OF SPIRAL GALAXIES AND DARK MATTER:

The forces balancing rotation of a mass "m" in an attractive gravitational field is:

(1) $(m * v * v)/r = M*m (Gn + A*r)/r*r$

where M is the attracting mass, m is the rotating mass, Gn is Newton's gravitational constant, r is the radius of rotation, and A is a coefficient in the first term in the series expansion of the gravitational constant.

The radius "ro" is defined as the radius where the Newtonian gravitational force is equal to the additional long-range force. The values of A and ro has initially been estimated and will be determined more accurately using curve fitting of rotational velocity as a function of r for various spiral galaxies.

Thus:

(2) $A = Gn/ro$

And:

(3) $(Gn + A*ro) = 2 * Gn$

When combined with equation 1, the following results:

(4) $v * v * ro = 2 * M * Gn$

This shows that for cases where the mass, M, of spiral galaxies is within an almost constant range of values, the outer rotational velocity, v, increases when the transition radius, ro, of the spiral galaxy is smaller. The radius, ro, is where the visible stars end, and where the rotational velocity becomes constant. Also, this shows that for spiral galaxies either the rotational velocity for those spiral galaxies increase slowly with the larger mass or radius ro of these spiral galaxies.

HUBBLE CONSTANT, VELOCITY, AND THE RAPIDLY EXPANDING UNIVERSE

According to our theory, the "optical Hubble constant", defined in terms of measured red shift and measured distance, and in units of red shift per unit of distance, can be shown to be larger for short distances and decrease for larger distances.

If one assumes that the red shift is a measure of the velocity, then one get a "velocity Hubble constant", in units of velocity per unit of distance, but for galaxies at large distances there is no experimental proof that the red shift can be used to measure velocity.

The potential energy well for a photon leaving the gravitation well of a star or galaxy and traveling a distance r is given by the integral of the attractive force with respect to the distance r.

The gravitational force under my theory is:

(5) $F = M*m \, (Gn/r*r + A/r)$

The loss of energy in moving away from the source of the gravitational force (which is a function of distance) is the force integrated over the distance of travel.

The loss of energy experienced by the photon results in a red shift that is proportional to the energy loss.

The energy loss and red shift can be described by

(6) $dE = M*m \, (Gn/r + A*\ln r)$

and when the term ln r is expanded in a power series in r, the following is first term of the result:

(7) $dE = k/r + s * r$

where k and s are constants.

The signs of each term are positive because it takes energy to move against attractive forces.

The term k/r corresponds to the energy to move out of a potential well, and the term s*r corresponds to the energy required to move against a long-range attractive force.

The red shift is proportional to the energy loss and the "optical Hubble constant" is proportional to the red shift divided by the distance.

(8)   Optical H = b (k/r*r + s)

Where "b", "k", and "s" are constants.

This analysis shows why the optical Hubble constant is large for small distances but decreases rapidly as the inverse square of the distance to an asymptotic constant value for large distances. With time, red shifts of further galaxies are added because of increased instrument sensitivity.

For very large Z values the power series should not be used in place of the ln(r) function otherwise the data may suggest faster, accelerating, expansion for very remote stars.

(9)   Optical H = b*k/r*r + s*b * ln(r)/r

REFERENCES AND READING MATERIAL

Y09 05 20 version

J. D. Anderson, P. Laing, P.A. Lau, E. L. Liu, A.S. Liu, M. M. Nieto, and S. G. Turyshev, Indication, from Pioneer 10/11, Galileo, and Ulysses Data, of an Apparent Anomalous, Weak, Long-Range Acceleration, Phys. Rev. Lett. 81, 14 (1998) pp. 2858-2861

H. Arp, Seeing Red: Red shifts, Cosmology and Academic Science, (Apeirion, Montreal, Canada, 1998). Also http:// red shift.vif.com

H. Arp, Quasars, Red shifts, and controversies, (Interstellar Media, 1987).

N. A. Bahcall, Large Scale Structure in the Universe", in Unsolved Problems in Astrophysics, edited by J. N.Bahcall, and J. P. Ostriker, (Princeton University Press, NJ, 1997), pp. 61-91.

G. Bothun, Modern Cosmological Observations and Problems, (Taylor & Francis, London, 1998). Other modifications of Newton's law have been proposed along with discussions of the many problems in the current cosmological models.

K. Ferguson. Measuring the Universe, (Walker and Company, N.Y., 1990).

D. Goldsmith, The Astronomers, (St. Martins Press, NY, 1991), pp. 36-44.

A. Guth, The Inflationary Universe, (Perseus Books, Cambridge Massachusetts, 1977).

E. R. Harrison, Cosmology: the Science of the Universe (Cambridge University Press, Cambridge, 1981), p. 240, discusses the tired light concept of Zwicky.

A. W. Hirschfield, Parallax, (W. H. Freeman, N.Y., 2001).

F. Hoyle, G. Burbidge, and J. V. Narlikar, A Different Approach to Cosmology, (Cambridge University Press, N.Y., 2000).

E. Hubble, A Relation between Distance and Radial Velocity among Extra Galactic Nebulae, Proceedings of the National Academy of Science, vol.15, pp. 168-73 (1929).

E. Hubble, The Observational Approach to Cosmology, (Clarendon Press, Oxford England, 1937) p. 68.

E. Hubble, and M. Humason, ApJ. 74, 43 (1931)

E. Hubble, The Realm of the Nebulae, (Yale University, New Haven 1936, 1982).

H. Kragh, Cosmology and Controversy, (Princeton University Press, Princeton, N.J. 1996).

R. P. Kushner, Extravagant Universe, (Princeton University, NJ, 2002).

E. J. Lerner, The Big Bang Never Happened, (Vintage Book, N.Y. c 1992).

M. Livo, The accelerating universe. (Jogn Wiley, 2000)

M. Milgrom, arXiv: astro-ph/9810302 v1 20 Oct 1998.

M. Milgrom, ApJ. Vol. 270, pp. 365-370. (1983)

M. Milgrom, Does Dark Matter Really Exist? Scientific American, pp.42-52, (2002).

P. J. E. Peebles, Principles of Physical Cosmology, (Princeton University Press, N.J.) 1993.

S. Perlmutter, Supernovae, Dark Energy, and the Accelerating Universe, Physics Today, April 2003, p. 53-60.

V. Rubin, and W. K. Ford, Astrophysics Journal 159:379 (1970).

W. C. Saslaw, Gravitational physics of stellar and galactic systems. Cambridge U. Press, Cambridge, 1985. (Viral theory)

J. Silk, The Big Bang, 3rd. ed., New York, W. H. Freedman and Company, 2001,

Y. Sofue and V. Rubin, Rotation Curves of Spiral Galaxies, in Annu. Rev. Astrophys. 2001. 39:137-74. Also http://www.physics.ucla.edu/~cwp/articles/rubindm/rubindm.html

L. Smolin, The life of the Cosmos, (Fordham University, N. Y., 1997)

D. Spergel, Dark Matter, in Unsolved Problems in Astrophysics, edited by J. M. Bahcall and J.P. Postriker, (Princeton University Press, Princeton, N.J., 1997) pp.221-240.

S. Weinberg, The First Three Minutes, second edition, (Basic Books, New York, 1988)

F. Zwicky, Red shift of Spectral Line, Proc. Nat. Acad. Sci., 1929, vol. 15, pp. 773-9

F. Zwicky, On the Possibilities of a Gravitational Drag of Light, Phys. Rev. Letters 34:Dec. 28, 1929.

Methods of gravity determination using laboratory techniques and equipment are discussed at http://mist.npl.washington.edu/eotwash/gconst.html

Evidence for the Big Bang - Remote Sensing Tutorial

http://rst.gsfc.nasa.gov/Sect20/A9.html

Techniques for measuring distances – describes 26 methods.

http://www.astro.ucla.edu/~wright/distance.htm

Sol Aisenberg, PHD
saisenberg@alum.mit.edu
solaisenberg@comcast.net
508/651-0140
Former: itgplus@earthlink.net

Draft 9.2.4
October 12, 2004